Lecture Notes in Mathematics

Edited by A. Dold and B. Eckmann

471

Richard S. Hamilton

T0222290

Harmonic Maps of Manifolds with Boundary

Springer-Verlag
Berlin · Heidelberg · New York 1975

Author
Prof. Richard S. Hamilton
Department of Mathematics
Cornell University
White Hall
Ithaca, N. Y. 14853
USA

Library of Congress Cataloging in Publication Data

Hamilton, Richard S 1943-
Harmonic maps of manifolds with boundary.

(Lecture notes in mathematics ; 471)
Bibliography: p.
Includes index.
1. Global analysis (Mathematics) 2. Manifolds (Mathematics) 3. Boundary value problems. 4. Function spaces. I. Title. II. Series: Lecture notes in mathematics (Berlin) ; 471.
QA3.L28 no.471 [QA614] 510'.8s [514'.223] 75-20001

AMS Subject Classifications (1970): 35J60, 35K55, 49A20, 49F15, 53C20, 58D15, 58E15, 58G99

ISBN 978-3-540-07185-3 Springer-Verlag Berlin · Heidelberg · New York
ISBN 978-0-387-07185-5 Springer-Verlag New York · Heidelberg · Berlin

Offsetdruck: Julius Beltz, Hemsbach/Bergstr.

TABLE OF CONTENTS

FOREWORD

The theory of harmonic maps of manifolds has its origins in the classic paper of Eells and Sampson [4], where existence is proved when the target manifold has non-positive Riemannian curvature. This paper generalizes this result to manifolds with boundary. Three results are stated corresponding to the Dirichlet, Neumann and mixed boundary value problems. The solution to the Dirichlet problem is proved in full detail, and we indicate the necessary minor modifications for the other two problems at the end. The paper is divided into five parts, and each part into sections.

Part I: Harmonic Maps

Part II: Function Spaces

Part III: Semi-Elliptic and Parabolic Equations

Part IV: The Heat Equation for Manifolds

Part V: Growth Estimates and Convergence.

In part I we define harmonic maps and state the three results. Part II contains the definition of weighted L^p spaces of potentials and Besov spaces used in the proof and reviews their properties. Part III reviews the theory of coercive linear semi-elliptic and parabolic boundary value problems. All the material in Parts II and III is well known to the experts in the field and can be found in the references in the bibliography. However, since we use several precise and delicate facts, which are scattered over many papers with many different definitions, and more particularly since the exposition of the subject has profited greatly from a recent multiplier theorem of Stein [22], we hope the reader will find this material a useful reference. The proof itself follows the method of Eells and Sampson in the construction of a non-linear heat equation for manifolds. In Part IV we prove uniqueness, regularity and existence for short periods of time. This material is independent of the curvature hypothesis, which first appears in Part V. Here we prove some rather delicate growth estimates which guarantee that the solution of the heat equation exists for all time and converges to a harmonic map.

The author would especially like to express his appreciation to James Eells Jr. for his invaluable advice and encouragement over many years, without which this paper would never have been written; and also to Halldor Eliasson, Ronald Goldstein, and Karen Uhlenbeck for many helpful discussions.

Cornell University
University of Warwick

Work partially supported by the SRC and NSF.

Part I: Harmonic Maps

1. Partial differential equations for maps $f:X \to Y$ of one manifold into another are of considerable interest in analysis and topology. In this context there are no linear equations, since Y has no additive structure. The polynomial equations of degree n are the simplest class of equations invariant under coordinate changes on X and Y. These are the equations given in local coordinates by polynomials in derivatives of f whose degrees sum to no more than n.

They look like

$$\sum_{|\alpha|+|\beta|+\ldots+|\gamma|\leq n} c_{\alpha\beta\ldots\gamma}(f)\; D^{\alpha}f D^{\beta}f \ldots D^{\gamma}f$$

where the coefficients $c_{\alpha\beta\ldots\gamma}(f)$ depend non-linearly on f and are multi-linear functionals applied to the vectors $D^{\alpha}f, D^{\beta}f, \ldots, D^{\gamma}f$. Here α denotes a multi-index $(\alpha_1, \ldots, \alpha_n)$ of length $|\alpha| = \alpha_1 + \ldots + \alpha_n$ and

$$D^{\alpha} = \left(\frac{\partial}{\partial x^1}\right)^{\alpha_1} \left(\frac{\partial}{\partial x^2}\right)^{\alpha_2} \cdots \left(\frac{\partial}{\partial x^n}\right)^{\alpha_n} .$$

2. The simplest and most important example is Laplace's equation, introduced for manifolds by Eells and Sampson [4]. Let X and Y be Riemannian manifolds with metrics g_{ij} and $h_{\alpha\beta}$, and $f:X \to Y$ a map between them. The derivative of f at a point $x \in X$ is a linear map $\nabla f_x : TX_x \to TY_{f(x)}$ on the tangent spaces. In the language of vector bundles, the derivative ∇f is a section of the bundle $L(TX, f^*TY)$ where f^*TY is the pull-back of TY to a bundle over X by the map f. In local coordinates

$$\nabla f = \frac{\partial f^{\alpha}}{\partial x^{i}}.$$

The second derivative $\nabla\nabla f$ is the derivative of ∇f with respect to the natural connection on $L(TX, f^*TY)$. This defines $\nabla\nabla f$ as a section of the bundle $L_s^2(TX, f^*TY)$ of symmetric bilinear maps. The Laplacian Δf is the trace of the second derivative $\nabla\nabla f$

$$\Delta f = \operatorname{Tr} \nabla\nabla f$$

with respect to the inner product on TX. This defines Δf as a section of the bundle f^*TY. In local coordinates, the Riemannian connections on TX and TY are given by the Christoffel symbols ${}_X\Gamma^i_{jk}$ and ${}_Y\Gamma^\alpha_{\beta\gamma}$. The pull-back connection on f^*TY is given by

$${}_Y\Gamma^\alpha_{\beta\gamma}(f)\,\frac{\partial f^\beta}{\partial x^i}\,.$$

where ${}_Y\Gamma^\alpha_{\beta\gamma}(f)$ is ${}_Y\Gamma^\alpha_{\beta\gamma}$ evaluated at $f(x)$.
If E and F are bundles, the connection on $L(E,F)$ is given in tensor analysis as minus the connection on E plus the connection on F. Thus in local coordinates

$$\nabla\nabla f = \frac{\partial^2 f^\alpha}{\partial x^i \partial x^j} - {}_X\Gamma^k_{ij}\,\frac{\partial f^\alpha}{\partial x^k} + {}_Y\Gamma^\alpha_{\beta\gamma}(f)\,\frac{\partial f^\beta}{\partial x^i}\,\frac{\partial f^\gamma}{\partial x^j}$$

and

$$\Delta f = g^{ij}\left\{\frac{\partial^2 f^\alpha}{\partial x^i \partial x^j} - {}_X\Gamma^k_{ij}\,\frac{\partial f^\alpha}{\partial x^k} + {}_Y\Gamma^\alpha_{\beta\gamma}(f)\,\frac{\partial f^\beta}{\partial x^i}\,\frac{\partial f^\gamma}{\partial x^j}\right\}.$$

The map $f:X \to Y$ is called harmonic if it satisfies Laplace's equation

$$\Delta f = 0.$$

This is the simplest elliptic second order polynomial partial differential equation for maps between manifolds.

3. There are many classical examples of harmonic maps.

(a) The harmonic maps $X \to R$ are the harmonic functions.

(b) The harmonic maps $R \to X$ are the geodesics.

(c) Every isometry is harmonic.

(d) A conformal map is one which preserves angles. Every conformal map is harmonic.

(e) Every holomorphic map between Kähler manifolds is harmonic.

(f) If $f:X_1 \times X_2 \to Y$ is harmonic in each variable separately then f is harmonic. In fact, there is a natural decomposition

$$\Delta f = \Delta_1 f + \Delta_2 f.$$

(g) If G is a Lie group with a bi-invariant Riemannian metric, then the multiplication $\mu:G\times G \to G$ is harmonic.

(h) The Hopf fibrations

$$S^3 \to S^2 \qquad S^7 \to S^4 \qquad S^{15} \to S^8$$

are harmonic in their classical polynomial representations.

(i) If Y is Riemannian and X is a submanifold of least volume, then the inclusion $i:X \to Y$ is harmonic for the induced metric on X.

4. The most important problem in the theory of harmonic maps is to prove or disprove the following conjecture. A homotopy class of maps of X into Y is a connected component of the space $\mathcal{M}(X,Y)$ of smooth maps of X into Y, with the C^∞ topology. Let X and Y be compact Riemannian manifolds without boundary.

Harmonic Conjecture: There exists a harmonic map in every homotopy class.

The best positive result is due to Eells and Sampson [4].

<u>Theorem</u>. If Y has Riemannian curvature ≤ 0 then there exists a harmonic map $f:X \to Y$ in every homotopy class.

The best negative result is due to Ted Smith [21]. He considers harmonic maps of a sphere into an ellipsoid of revolution which are of degree k and axially symmetric. These exist if the ellipsoid is short and fat, but not if it is tall and thin. Thus as the ellipsoid becomes taller and thinner, at some point the harmonic map either bifurcates into a family of axially asymmetric maps, or it ceases to exist at all. Which happens is not known.

5. In this paper we extend the result of Eells and Sampson to compact manifolds X and Y with boundary. There are three natural boundary value problems.

(a) Dirichlet Problem.

We ask for a harmonic map $f:X \to Y$ with given values on ∂X. Let $h:\partial X \to Y$ be a smooth map of ∂X into Y. Let $\mathcal{M}_h(X,Y)$ denote the closed subspace of maps $f:X \to Y$ with $f|\partial X = h$. A relative homotopy class is a connected component of $\mathcal{M}_h(X,Y)$. If there is a topological obstruction to extending h then $\mathcal{M}_h(X,Y)$ is empty and nothing more can be said. Otherwise we have the following theorem.

<u>Theorem</u>. Let X and Y be compact Riemannian manifolds with boundary. Suppose that Y has Riemannian curvature ≤ 0 and that ∂Y is convex (or empty). Then the Dirichlet problem for $f:X \to Y$

$$\Delta f = 0 \quad \text{on} \quad X$$
$$f = h \quad \text{on} \quad \partial X$$

has a solution in every relative homotopy class.

The condition that ∂Y is convex is a local condition which can be expressed in terms of the Christoffel symbols. Choose a chart $(y^1,\ldots,y^{n-1},y^n)$ near ∂Y such that $Y = \{y^n \geq 0\}$. The condition that ∂Y is convex is that in such a chart the matrix $\Gamma^n_{\alpha\beta}(1 \leq \alpha,\beta \leq n-1)$ is (weakly) positive definite. To see the geometric meaning consider a geodesic $\varphi = \varphi^{\alpha}(t)$ passing through a point on ∂Y. The equation for a geodesic says

$$\frac{d^2\varphi^n}{dt^2} + \Gamma^n_{\alpha\beta} \frac{d\varphi^\alpha}{dt} \frac{d\varphi^\beta}{dt} = 0.$$

If φ is tangent to ∂Y, $\frac{d\varphi^n}{dt} = 0$ and only terms $\Gamma^n_{\alpha\beta}$ with $1 \leq \alpha,\beta \leq n-1$ appear. If $\Gamma^n_{\alpha\beta}$ $(1 \leq \alpha,\beta \leq n-1)$ is positive definite then $\frac{d^2\varphi^n}{dt^2} \leq 0$. Thus the condition that ∂Y is convex is that a geodesic tangent to ∂Y does not enter inside Y. If $X = R$ then the harmonic maps are the geodesics, so the condition that ∂Y is convex is clearly necessary.

(b) Neumann Problem.

If we do not specify the map f on ∂X at all, we can impose instead the auxiliary condition that the normal derivative $\nabla_\nu f = 0$ on ∂X. Note that $\nabla f_x : TX_x \to TY_{f(x)}$ and $\nu \in TX_x$ is the normal vector, so $\nabla_\nu f = \nabla f(\nu) \in TY_{f(x)}$ is a tangent vector on Y.

<u>Theorem</u>. Let X and Y be compcat Riemannian manifolds with boundary. Suppose that Y has Riemannian curvature ≤ 0 and that ∂Y is convex (or empty). Then the Neumann Problem

$$\Delta f = 0 \quad \text{on } X$$
$$\nabla_\nu f = 0 \quad \text{on } \partial X$$

has a solution in every homotopy class.

(c) Mixed Problem.

The two preceding problems do not involve ∂Y in an essential way. This one does. Suppose that we require that f maps ∂X into ∂Y, but in an arbitrary fashion. We can then impose the auxiliary boundary condition that the normal derivative $\nabla_\nu f$ taken at a point in ∂X should be normal to ∂Y. This makes sense since

$\nabla_\nu f_x \in TY_{f(x)}$. Let $\mathcal{M}_\partial(X,Y)$ denote the closed subspace of those $f \in \mathcal{M}(X,Y)$ with $f(\partial X) \subseteq \partial Y$. A relative homotopy class will now mean a connected component of $\mathcal{M}_\partial(X,Y)$. To prove a theorem in this case we must assume more about ∂Y, namely that it is totally geodesic. Hopefully this requirement may be weakened. In a local chart the condition that ∂Y be totally geodesic is that the matrix $\Gamma^n_{\alpha\beta}(1 \leq \alpha,\beta \leq n-1)$ discussed before should be zero. This condition says that a geodesic tangent to ∂Y lies entirely inside ∂Y.

Theorem. Let X and Y be compact Riemannian manifolds with boundary. Suppose that Y has Riemannian curvature ≤ 0 and that ∂Y is totally geodesic. Then the mixed problem

$$\Delta f = 0 \quad \text{on } X$$
$$f(\partial X) \subseteq \partial Y$$
$$\nabla_\nu f \perp \partial Y \quad \text{on } \partial X$$

has a solution in every relative homotopy class.

Part II: Function Spaces

1. The Fourier transform is the basic tool for studying constant coefficient partial differential equations. Let X be the n-dimensional Euclidean space of the variables $\{x_1,\ldots,x_n\}$ and let Ξ be the dual space with variables $\{\xi_1,\ldots,\xi_n\}$, under the pairing

$$\langle x,\xi\rangle = x_1\xi_1+\cdots x_n\xi_n.$$

Let $\mathcal{S}(X)$ be the space of smooth rapidly decreasing functions on X. If $x^\alpha = x_1^{\alpha_1}\cdots x_n^{\alpha_n}$ and

$$D^\alpha = \left(\frac{1}{i}\frac{\partial}{\partial x_1}\right)^{\alpha_1}\left(\frac{1}{i}\frac{\partial}{\partial x_2}\right)^{\alpha_2}\cdots\left(\frac{1}{i}\frac{\partial}{\partial x_n}\right)^{\alpha_n}$$

then

$$\mathcal{S}(X) = \{f\colon x^\alpha D^\beta f \to 0 \quad \text{as} \quad x\to\infty \quad \text{for all} \quad \alpha,\beta\}.$$

Then the Fourier transform

$$\hat{f}(\xi) = \int_X e^{-i\langle\xi,x\rangle} f(x)\,dx$$

defines an isomorphism of $\mathcal{S}(X)$ onto $\mathcal{S}(\Xi)$. This gives a decomposition of f into plane waves

$$f(x) = \int_\Xi e^{i\langle\xi,x\rangle}\hat{f}(\xi)\,d\xi.$$

If $P(D) = \Sigma c_\alpha D^\alpha$ is a partial differential operator with constant coefficients

$$\widehat{P(D)f} = P(\xi)\hat{f}(\xi)$$

where $P(\xi) = \Sigma c_\alpha \xi^\alpha$.

This generalizes to the case where P is not a polynomial. Let $\mathcal{M}(\Xi)$ denote the space of smooth slowly growing functions on Ξ; that is, those functions $M(\xi)$ such that every derivative $D^\alpha M(\xi)$ grows slower than some polynomial in ξ as $\xi\to\infty$.

Multiplication defines a map

$$\mathcal{M}(\Xi) \times \mathcal{S}(\Xi) \to \mathcal{S}(\Xi).$$

We define the operator $M(D)\colon \mathcal{S}(X) \to \mathcal{S}(X)$ by the rule

$$\widehat{M(D)f} = M(\xi)\hat{f}(\xi).$$

2. Let $L^p(X)$ denote the space of measurable p-integrable functions on X, with the usual norm

$$\|f\|_{L^p} = \{\int_X |f(x)|^p dx\}^{1/p}.$$

The theory of Fourier multipliers deals with the question of when an operator $M(D)\colon \mathcal{S}(X) \to \mathcal{S}(X)$ extends by completion to a continuous map $M(D)\colon L^p(X) \to L^p(X)$. A very general theorem to this effect has been proved by Elias Stein in [22]. It has the following simpler result as a corollary. An index α is primitive if $\alpha = (\alpha_1, \ldots, \alpha_n)$ with each α_i either 0 or 1. There are 2^n primitive indices.

Theorem (Stein): Suppose that for each primitive index α there is a constant C_α with

$$|\xi^\alpha D^\alpha m(\xi)| \leq C_\alpha.$$

Then $M(D)$ defines a continuous linear map of $L^p(X)$ into itself when $1 < p < \infty$. Moreover $\|M(D)f\|_{L^p} \leq C\|f\|_{L^p}$ with a constant C depending linearly only on the C_α.

3. Motivated by this we make the following definition.

<u>Definition</u>. A multiplier $W(\xi) \in \mathcal{M}(\Xi)$ is called a weight function if $W(\xi) > 0$ and for every primitive index α there is a constant C_α with

$$|\xi^\alpha D^\alpha W(\xi)| \leq C_\alpha W(\xi).$$

It is easy to verify that if $W(\xi)$ and $V(\xi)$ are weight functions, so are $W(\xi)+V(\xi)$, $W(\xi)\ V(\xi)$, and $W(\xi)^s$ for any real s. We write $W(D)^s$ for the operator corresponding to the multiplier $W(\xi)^s$.

The space of temperate distributions $\mathcal{S}(X)^*$ is the topological vector space of continuous linear functionals on $\mathcal{S}(X)$, with the strong topology of uniform convergence on bounded subsets. There are natural dense inclusions $\mathcal{S}(X) \subseteq L^p(X) \subseteq \mathcal{S}(X)^*$. The operators $M(D)\colon \mathcal{S}(X) \to \mathcal{S}(X)$ extend by duality to continuous operators $M(D)\colon \mathcal{S}(X)^* \to \mathcal{S}(X)^*$.

Definition. For any real s

$$L^p_s(X;W) = \{f \in \mathcal{S}(X)^*\colon W(D)^s f \in L^p(X)\}.$$

This is a Banach space with the norm

$$\|f\|_{L^p_s} = \|W(D)^s f\|_{L^p}.$$

For any real r

$$W(D)^r\colon L^p_{r+s}(X;W) \to L^p_s(X;W)$$

is continuous. If $V(\xi) \leq CW(\xi)$ then there is a continuous inclusion $L^p_s(X;W) \subseteq L^p_s(X;V)$, since Stein's theorem applies to $V(\xi)^s W(\xi)^{-s}$.

Only one kind of weight function will interest us. In semi-elliptic equations we assign each variable x_i an integer weight σ_i, and count σ_i derivatives in the x_i direction with the same weight as σ_j derivatives in the x_j direction. Let $\Sigma = (\sigma_1,\ldots,\sigma_n)$ be such a collection of integers and let σ be their least common multiple. Define the weight function

$$W_\Sigma(\xi) = (1 + \xi_1^{2\sigma_1} + \cdots + \xi_n^{2\sigma_n})^{1/2\sigma}.$$

It is easy to check that $|\xi^{\alpha}D^{\alpha}W_{\Sigma}(\xi)| \leq C_{\alpha}W_{\Sigma}(\xi)$ for all indices α. We shall write $L^p_s(X)$ for $L^p_s(X;W_{\Sigma})$ unless it is necessary to specify the weight function. As an example, for the heat operator

$$\frac{\partial}{\partial t} - \frac{\partial^2}{\partial x_1^2} - \cdots - \frac{\partial^2}{\partial x_n^2}$$

we assign weight 2 to the variables $x_1,\ldots,x_n$ and weight 1 to the variable t.

We can give an explicit description of the spaces $L^p_k(X;W_{\Sigma})$ when k is an integer multiple of σ. Let

$$\|\alpha\| = \sigma\left(\frac{\alpha_1}{\sigma_1} + \cdots + \frac{\alpha_n}{\sigma_n}\right)$$

define the weight of the index α.

<u>Theorem</u>. Let k be a positive integer multiple of σ. Then for $1 < p < \infty$

(1) $f \in L^p_k(X;W_{\Sigma}) \Leftrightarrow D^{\alpha}f \in L^p(X)$ when $\|\alpha\| \leq k$

and

$$\sum_{\|\alpha\| \leq k} \|D^{\alpha}f\|_{L^p(X)} \text{ is equivalent to } \|f\|_{L^p_k(X;W)}$$

(2) $f \in L^p_{-k}(X;W_{\Sigma}) \Leftrightarrow \exists g_{\alpha} \in L^p(X)$ with $f = \sum_{\|\alpha\| \leq k} D^{\alpha}g_{\alpha}$

and

$$\inf\left\{ \sum_{\|\alpha\| \leq k} \|g_{\alpha}\|_{L^p(X)} : f = \sum_{\|\alpha\| \leq k} D^{\alpha}g_{\alpha} \right\}$$

is equivalent to $\|f\|_{L^p_{-k}(X;W)}$.

<u>Proof</u>. $W^{\sigma}_{\Sigma}(\xi) = (1 + \xi_1^{2\sigma_1} + \cdots + \xi_n^{2\sigma_n})^{1/2\sigma}$. Let

$$E_0(\xi) = 1/W^{\sigma}_{\Sigma}(\xi)$$

$$E_1(\xi) = \xi_1^{\sigma_1}/W^{\sigma}_{\Sigma}(\xi),\ldots,E_n(\xi) = \xi_n^{\sigma_n}/W^{\sigma}_{\Sigma}(\xi).$$

Then $W^{\sigma}_{\Sigma}(\xi) = E_0(\xi) + \xi_1^{\sigma_1}E_1(\xi) + \cdots + \xi_n^{\sigma_n}E_n(\xi)$.

Thus $W^{\sigma}_{\Sigma}(D) = E_0(D) + D_1^{\sigma_1}E_1(D) + \cdots + D_n^{\sigma_n}E_n(D)$.

Moreover by Stein's multiplier theorem we know that $E_0(D), E_1(D), \ldots, E_n(D)$ give bounded linear maps of $L^p(X)$ to itself for $1 < p < \infty$. Hence if $k = m\sigma$ is a positive integer multiple of σ we can write

$$W_\Sigma^k(D) = \sum_{\|\alpha\| \leq k} E_\alpha(D) \cdot D^\alpha$$

where the $E_\alpha(D)$ are bounded linear maps on $L^p(X)$. Conversely it follows from Stein's theorem that if $F_\alpha(\xi) = \xi^\alpha / W_\Sigma^k(\xi)$ then, for $\|\alpha\| \leq k$, $F_\alpha(D)$ is a bounded linear map on $L^p(X)$, and

$$D^\alpha = F_\alpha(D)\, W_\Sigma^k(D).$$

The rest of the proof is obvious.

4. The next major tool is the theory of holomorphic interpolation of Banach spaces as developed by Calderón in [2]. It is based on the classical Three Lines Theorem of complex analysis (see Rudin [20], pg. 243).

Three Lines Theorem. Let $h(z)$ be continuous on the strip $S = \{0 \leq \operatorname{Re} z \leq 1\}$, converging to 0 at ∞ and holomorphic in the interior. Let $z = x + iy$ and

$$M(x) = \sup_y |h(x+iy)|.$$

Then $M(x) \leq M(1)^x M(0)^{1-x}$ for $0 \leq x \leq 1$.

Let Ω be an open set in the complex plane and $h: \Omega \to A$ a continuous map of Ω into a complex Banach space A. We say h is holomorphic if any of the following three equivalent conditions hold:

(a) h is Fréchet differentiable and the derivative is complex-linear at each point.

(b) The Cauchy integral formula is valid.

(c) The composition of h with any continuous linear functional on A is holomorphic in the ordinary sense.

The preceding theorem generalizes as follows.

Theorem. Let $h(z)$ be continuous on the strip $\{0 \leq \operatorname{Re} z \leq 1\}$ with values in a complex Banach space A, converging to 0 at ∞ and holomorphic in the interior. Let

$$M(x) = \sup_y \| h(x+iy) \| .$$

Then $M(x) \leq M(1)^x M(0)^{1-x}$.

Proof. Let λ be any continuous linear functional on A with $\|\lambda\| \leq 1$. Let

$$M_\lambda(x) = \sup_y \| \lambda \circ h(x+iy) \| .$$

Since $\lambda \circ h$ is an ordinary holomorphic function

$$M_\lambda(x) \leq M_\lambda(1)^x M_\lambda(0)^{1-x}.$$

Then $M(x) = \sup\{M_\lambda(x) : \|\lambda\| \leq 1\}$ so

$$M(x) \leq M(1)^x M(0)^{1-x}.$$

We say that two complex Banach spaces A_0 and A_1 form an interpolation pair if $A_0 \cap A_1$ is a dense linear subspace of A_0 and also of A_1, with the property that if a_n is a sequence in $A_0 \cap A_1$ which converges to a in A_0 and also converges to a' in A_1, then $a = a'$. This will be the case if A_0 and A_1 contain a common dense subspace $\mathcal{D}$, and are contained in an enveloping Hausdorff topological vector space $\mathcal{E}$, with continuous inclusions. Then $A_0 \cap A_1$ is a Banach space with norm

$$\| a \|_{A_0 \cap A_1} = \| a \|_{A_0} + \| a \|_{A_1}$$

and $A_0 + A_1$ is a Banach space with norm

$$\|a\|_{A_0+A_1} = \inf\{\|a_0\|_{A_0} + \|a_1\|_{A_1} : a_0 + a_1 = a\}.$$

There is an obvious exact sequence

$$0 \to A_0 \cap A_1 \to A_0 \oplus A_1 \to A_0 + A_1 \to 0.$$

Let $\mathcal{H}(A_0, A_1)$ denote the Banach space of all functions $h(z)$ defined on the strip $S = \{0 \leq \operatorname{Re} z \leq 1\}$ with values in $A_0 + A_1$ such that

1) h is a bounded continuous map of $\{\operatorname{Re} z = 0\}$ into A_0 converging to 0 at ∞.

2) h is a bounded continuous map of $\{\operatorname{Re} z = 1\}$ into A_1 converging to 0 at ∞.

3) h is a bounded continuous map of $\{0 \leq \operatorname{Re} z \leq 1\}$ into $A_0 + A_1$ converging to 0 at ∞ which is holomorphic in the interior.

We take $\|h\|_{\mathcal{H}(A_0,A_1)} = M_0 + M_1$ where

$$M_0 = \sup\{\|h(z)\|_{A_0} : \operatorname{Re} z = 0\}$$

$$M_1 = \sup\{\|h(z)\|_{A_1} : \operatorname{Re} z = 1\}.$$

If we also put

$$M(x) = \sup\{\|h(z)\|_{A_0+A_1} : \operatorname{Re} z = x\}$$

then by the previous theorem

$$M(x) \leq M(1)^x M(0)^{1-x} \leq M_0^x M_1^{1-x}$$

so the bounds on the sides of the strip give a bound in the interior. We define for $0 \leq \theta \leq 1$

$$\mathcal{K}_\theta(A_0, A_1) = \{h \in \mathcal{H}(A_0, A_1) : h(\theta) = 0\}$$

and $A_\theta = \mathcal{H}(A_0, A_1)/\mathcal{K}_\theta(A_0, A_1)$ with the quotient norm. Thus

$$A_\theta = \{h(\theta) : h \in \mathcal{H}(A_0, A_1)\}$$

$$\| a \|_{A_\theta} = \inf\{M_0 + M_1 : h(\theta) = a\}.$$

We immediately remark that if $\theta = 0$ or 1 we recover the space A_0 or A_1. Thus if $\theta = 0$ and $a \in A_0$, we can approximate a by elements $a_n \in A_0 \cap A_1$ and consider the functions $h_{n,k}(z) = e^{-kz} a_n$. Then $\| h_{n,k} \|_{\mathcal{H}(A_0,A_1)} = \| a_n \|_{A_0} + e^{-k} \| a_n \|_{A_1}$. Thus $h_{n,k}(\theta) \to a$ and $\| h_{n,k} \| \to \| a \|_{A_0}$ as $n,k \to \infty$. The same holds for $\theta = 1$.

<u>Theorem</u>. If $h \in \mathcal{H}(A_0, A_1)$ then

$$\| h(\theta) \|_{A_\theta} \leq 2\, M_1^\theta M_0^{1-\theta} .$$

<u>Proof</u>. Replace $h(z)$ by

$$\tilde{h}(z) = e^{c(z-\theta)} h(z).$$

Then $\tilde{M}_0 = e^{-c\theta} M_0$ and $\tilde{M}_1 = e^{c(1-\theta)} M_1$ while $\tilde{h}(\theta) = h(\theta)$. Thus

$$\| h(\theta) \|_{A_\theta} = \| \tilde{h}(\theta) \|_{A_\theta} \leq \tilde{M}_0 + \tilde{M}_1$$

$$= e^{-c\theta} M_0 + e^{c(1-\theta)} M_1 \leq 2\, M_1^\theta M_0^{1-\theta}$$

if c is chosen with $e^c = M_0/M_1$.

<u>Corollary</u>. If $a \in A_0 \cap A_1$ then

$$\| a \|_{A_\theta} \leq 2 \| a \|_{A_1}^{\theta} \| a \|_{A_0}^{1-\theta}.$$

<u>Proof</u>. Take $h(z) = a$ for all z.

Suppose now that (A_0, A_1) is one interpolation pair of Banach spaces as before and that (B_0, B_1) is another such pair. Suppose that E defines a continuous linear map of A_0 into B_0

and also of A_1 into B_1, such that the definitions agree on $A_0 \cap A_1$. Then using the exact sequence $0 \to A_0 \cap A_1 \to A_0 \oplus A_1 \to A_0 + A_1 \to 0$ it follows that E defines a continuous linear map of $A_0 + A_1$ into $B_0 + B_1$, agreeing with the original definitions on A_0 and A_1, as can be seen from the diagram

$$\begin{array}{ccccccccc} 0 \to & A_0 \cap A_1 & \to & A_0 \oplus A_1 & \to & A_0 + A_1 & \to 0 \\ & \downarrow E & & \downarrow E\oplus E & & \downarrow E & \\ 0 \to & B_0 \cap B_1 & \to & B_0 \oplus B_1 & \to & B_0 + B_1 & \to 0 \end{array}$$

Theorem. E defines a continuous linear map of A_θ into B_θ for $0 \leq \theta \leq 1$ and

$$\| E\|_{L(A_\theta,B_\theta)} \leq 2\| E\|^{\theta}_{L(A_1,B_1)} \| E\|^{1-\theta}_{L(A_0,B_0)} .$$

Proof. It is immediate that E defines a continuous linear map of $\mathcal{H}(A_0,A_1)$ into $\mathcal{H}(B_0,B_1)$, by $(Eh)(z) = E(h(z))$. Since E takes $\mathcal{K}_\theta(A_0,A_1)$ into $\mathcal{K}_\theta(B_0,B_1)$, upon passing to the quotients E defines a continuous linear map of A_θ into B_θ. Finally, if $a \in A_\theta$ we can find, for each $\varepsilon > 0$, a function $h \in \mathcal{H}(A_0,A_1)$ with $h(\theta) = a$ and $\| h\|_{\mathcal{H}(A_0,A_1)} = M_0 + M_1 \leq (1+\varepsilon)\| a\|_{A_\theta}$ where we recall

$$M_0 = \sup\{\| h(z)\|_{A_0} : \text{Re } z = 0\}$$

$$M_1 = \sup\{\| h(z)\|_{A_1} : \text{Re } z = 1\}.$$

Let $\tilde{h} = E\circ h$. Then $\tilde{M}_0 \leq C_0M_0$ and $\tilde{M}_1 \leq C_1M_1$ where $C_0 = \| E\|_{L(A_0,B_0)}$ and $C_1 = \| E\|_{L(A_1,B_1)}$. Then $E\circ h(\theta) = Ea$ so

$$\| Ea\|_{B_\theta} \leq 2(C_1M_1)^{\theta}(C_0M_0)^{1-\theta} \leq 2\, C_1^{\theta}C_0^{1-\theta}(1+\varepsilon)\| a\|_{\theta}.$$

But $\varepsilon > 0$ is arbitrary, so

$$\| E\|_{L(A_\theta,B_\theta)} \leq 2\| E\|^{\theta}_{L(A_1,B_1)} \| E\|^{1-\theta}_{L(A_0,B_0)} .$$

5. Our interest in holomorphic interpolation lies in the following result.

Theorem. Let $1 < p,q < \infty$ and let W be any weight function. If $A_0 = L^q_m(X;W)$ and $A_1 = L^p_n(X;W)$ then the interpolation space $A_\theta = L^r_k(X;W)$ with an equivalent norm where $\frac{\theta}{p} + \frac{1-\theta}{q} = \frac{1}{r}$ and $\theta n + (1-\theta)m = k$.

Proof. To begin we choose two linear functions $\lambda(z)$ and $\mu(z)$ so that

$$\lambda(0) = 1/q \qquad \lambda(\theta) = 1/r \qquad \lambda(1) = 1/p$$
$$\mu(0) = m \qquad \mu(\theta) = k \qquad \mu(1) = n.$$

Lemma (a). Let $f \in L^r$. For every $\delta > 0$ we can find smooth functions φ and ψ with compact support such that

$$\|\varphi\|_{L^\infty} \leq 1 \qquad \|\psi\|_{L^1} \leq \|f\|^r_{L^r} \qquad \|f - \varphi\psi^{1/r}\|_{L^r} \leq \delta\|f\|_{L^r}.$$

Proof. It is enough to prove the lemma when f itself is smooth with compact support, for these are dense in L^r. If $f = 0$ we can take $\varphi = 0$, $\psi = 0$. Otherwise choose $\epsilon > 0$ so that, if f has support in a ball of volume V, we have $\epsilon \cdot V^{1/r} \leq \delta\|f\|_{L^r}$. Next choose a smooth function $\rho(z)$ with $\rho(z) > 0$ for $z \neq 0$ and $\rho(z) = e^{-1/|z|}$ for small z so that

$$|z| - \epsilon \leq \rho(z) \leq |z|.$$

The choice of $\rho(z)$ near $z = 0$ guarantees that $\rho(z)^r$ is smooth also. Let

$$\sigma(z) = \frac{z}{\epsilon + \rho(z)}.$$

Then $\sigma(z)$ is smooth and $|\sigma(z)| \leq 1$. Also

$$|z - \rho(z)\sigma(z)| = \left|\frac{\epsilon z}{\epsilon + \rho(z)}\right| \leq \epsilon.$$

The idea here is that $\rho(z)$ and $\sigma(z)$ are smooth approximations to $|z|$ and $\arg z$. Let $\psi = \rho(f)^r$ and $\varphi = \sigma(f)$. Then both are smooth with compact support and

$$\|\varphi\|_{L^\infty} \leq 1 \qquad \|\psi\|_{L^1} \leq \|f\|^r_{L^r}.$$

$$\|f-\varphi\psi^{1/r}\|_{L^r} \leq \varepsilon\cdot V^{1/r} \leq \delta\|f\|_{L^r}.$$

Lemma (b). Let E and G be Banach spaces and $\lambda\colon G \to E$ a continuous linear map. Suppose there exists a $\delta < 1$ and a constant C such that for every $x \in E$ we can find $y \in G$ with

$$\|y\|_G \leq C\|x\|_E \quad \text{and} \quad \|x-\lambda y\|_E \leq \delta\|x\|_E.$$

Then λ maps G onto E.

Proof. Let x_0 be any point of E. Choose inductively $y_0,x_1,y_1,x_2,y_2,\ldots,x_n,y_n$ so that

$$y_n \in G \qquad \|y_n\|_G \leq C\|x_n\|_E \qquad \|x_n-\lambda y_n\|_E \leq \delta\|x_n\|_E$$

$$x_{n+1} \in E \qquad x_{n+1} = x_n - \lambda y_n.$$

Then $\|x_{n+1}\|_E \leq \delta\|x_n\|_E$ so $\|x_n\|_E \leq \delta^n\|x_0\|_E$ and $\|y_n\|_G \leq C\delta^n\|x_0\|_E$. Thus $x_n \to 0$ in E and Σy_n converges absolutely in G. Now $x_n = \lambda y_n + x_{n+1}$ so

$$x_0 = \lambda y_0 + x_1 = \lambda y_0 + \lambda y_1 + x_2 = \ldots$$

$$x_0 = \lambda(y_0+y_1+\ldots+y_n) + x_{n+1}$$

As $n \to \infty$, $y_0+y_1+\ldots+y_n \to \Sigma y_n$ and $x_{n+1} \to 0$. Thus $x_0 = \lambda(\Sigma y_n)$. Hence λ maps G onto E.

Lemma (c). If iy is purely imaginary,

$$\|W^{iy}f\|_{L^r} \leq C(1+|y|)^n\|f\|_{L^r}$$

for $1 < r < \infty$ and any weight function W.

Proof. This follows directly from Stein's multiplier theorem. In order to estimate the bound on the multiplier we only have to estimate $|\xi^\alpha D^\alpha W^{iy}(\xi)|$ for primitive α. Typical derivatives are

$$\xi_1 \frac{\partial}{\partial \xi_1} W^{iy}(\xi) = iy(\xi_1 \frac{\partial W}{\partial \xi_1}) W^{iy-1}(\xi)$$

$$\xi_1 \xi_2 \frac{\partial^2}{\partial \xi_1 \partial \xi_2} W^{iy}(\xi) = iy(\xi_1 \xi_2 \frac{\partial^2 W}{\partial \xi_1 \partial \xi_2}) W^{iy-1}(\xi)$$

$$+ iy(iy-1)(\xi_1 \frac{\partial W}{\partial \xi_1})(\xi_2 \frac{\partial W}{\partial \xi_2}) W^{iy-2}(\xi)$$

and so on. The assumption that W is a weight function is that $W > 0$ and

$$\xi_1 \frac{\partial W}{\partial \xi_1} \leq CW, \quad \xi_2 \frac{\partial W}{\partial \xi_2} \leq CW, \quad \xi_1 \xi_2 \frac{\partial^2 W}{\partial \xi_1 \partial \xi_2} \leq CW$$

and so on. Finally $|W^{iy}| = 1$. Thus all the quantities to be estimated are bounded by polynomials in y of degree no more than n (the dimension of the space).

Lemma (d). If φ and ψ are smooth with compact support then

$$h(z) = e^{(z-\theta)^2} W^{-\mu(z)}(D)(\varphi \cdot \psi^{\lambda(z)})$$

is a continuous and bounded map of the strip $\{0 \leq \text{Re } z \leq 1\}$ into $\mathcal{S}(X)$ and is holomorphic in the interior converging to 0 at ∞.

Proof. Write $z = x+iy$; then $0 \leq x \leq 1$ in the strip and $\lambda(z) = \lambda(x) + i\lambda(y)$, $\mu(z) = \mu(x) + i\mu(y)$ are the real and imaginary parts. Also $|e^{(z-\theta)^2}| \leq Ce^{-y^2}$ goes to zero faster than any polynomial in y.

The topology on $\mathcal{S}(X)$ is defined by norms

$$\|f\|_{n,k} = \sum_{|\beta| \leq k} \sup(1+|x|^2)^{n/2} |D^\beta f(x)|.$$

The bounded sets in $\mathcal{S}(X)$ are the sets B such that

$$\forall n \quad \forall k \quad \exists C \quad \forall f \in B \quad \|f\|_{n,k} \leq C.$$

The function $z \to \varphi \cdot \psi^{\lambda(z)} \in \mathcal{S}(X)$ is continuous on the strip and holomorphic in the interior; and since each x-derivative introduces another coefficient $[\lambda(z)-j]$ with integer j, we have

$$\forall m \quad \forall k \quad \exists C$$

$$\| \varphi \cdot \psi^{\lambda(z)} \|_{m,k} \leq C(1+|y|)^k.$$

The Fourier transform defines an isomorphism of $\mathcal{S}(X)$ onto $\mathcal{S}(\Xi)$, and if $s > \dim X$,

$$\|\hat{f}\|_{n,k} \leq C\|f\|_{k+s,n}.$$

The topology on the space $\mathcal{M}(\Xi)$ of smooth slowly growing functions of ξ is given by pseudo-norms

$$\|m\|_{-n,k} = \sum_{|\beta|\leq k} \sup(1+|\xi|^2)^{-n/2}|D^{\beta}m(\xi)|$$

where $\mathcal{M}(\Xi) = \{m\colon \forall k \ \exists n \ \ \|m\|_{-n,k} < \infty\}$. The bounded sets of $\mathcal{M}(\Xi)$ are the sets B such that

$$\forall k \quad \exists n \quad \exists C \quad \forall m \in B \quad \|m\|_{-n,k} \leq C.$$

For any multiplier $W(\xi) > 0$ the map $z \to W(\xi)^{-\mu(z)}$ is continuous on the strip and holomorphic in the interior; and we have the estimate

$$\forall k \quad \exists n \quad \exists C$$

$$\|W(\xi)^{-\mu(z)}\|_{-n,k} \leq C(1+|y|)^k.$$

The bilinear map

$$\mathcal{M}(\Xi) \times \mathcal{S}(X) \to \mathcal{S}(X)$$

given by $m(\xi) \times f(x) \to m(D)f(x)$ is jointly continuous; in fact $\forall n,k,r \quad \exists C$ such that

$$\|m(\xi)\hat{f}(\xi))\|_{n,k} \leq C\|m(\xi)\|_{-r,k}\|\hat{f}\|_{n+r,k}$$

so if $s > \dim X$

$$\|m(D)f(x)\|_{n,k} \leq C\|m(\xi)\|_{-r,n}\|f(x)\|_{n+s,k+r+s}.$$

Therefore the map $z \to W(\xi)^{-\mu(z)}(\varphi\cdot\psi^{\lambda(z)})$ is continuous on the strip and holomorphic in the interior, and satisfies the estimate

$$\forall n \quad \forall k \quad \exists r \quad \exists C$$

$$\|W(D)^{-\mu(z)}(\varphi\cdot\psi^{\lambda(z)})\|_{n,k} \leq C(1+|y|)^{k+r+s}.$$

Since $e^{(z-\theta)^2}$ goes to zero faster than any power of y, we have

$\forall n,k \quad \exists C \quad \|h(z)\|_{n,k} \leq C$ so $h(z)$ is bounded in $\mathcal{S}(X)$.

Now we can complete the proof of the theorem. For every $g \in L_k^r(X;W)$ let $f = W^k g \in L^r(X)$. Choose φ and ψ smooth with compact support as in Lemma (a), so that, for a fixed $\delta < 1$,

$$\|\varphi\|_{L^\infty} \leq 1 \quad \|\psi\|_{L^1} \leq \|f\|^r_{L^r} \quad \|f-\varphi\psi^{1/r}\|_{L^r} \leq \delta\|f\|_{L^r}.$$

Let

$$h(z) = e^{(z-\theta)^2} W^{-\mu(z)}(D)(\varphi\cdot\psi^{\lambda(z)}).$$

By Lemma (d) we certainly have $h \in \mathcal{H}(A_0,A_1)$ and

$$h(\theta) = W^{-k}(D)(\varphi\cdot\psi^{1/r})$$

so

$$\|g-h(\theta)\|_{L_k^r(X;W)} \leq \delta\|g\|_{L_k^r(X;W)}.$$

Moreover, using Lemma (c) we have

$$M_0 = \sup\{\|h(z)\|_{L_m^q(X;W)} : \text{Re } z = 0\} \leq C\|\psi^{1/q}\|_{L^q}$$

$$M_0 \leq C\|f\|^{r/q}_{L^r}$$

$$M_1 = \sup\{\|h(z)\|_{L_n^p(X;W)} : \text{Re } z = 1\} \leq C\|\psi^{1/p}\|_{L^p}$$

$$M_1 \leq C\|f\|^{r/p}_{L^r}.$$

Therefore

$$\|h(\theta)\|_{A_\theta} \leq CM_1^\theta M_0^{1-\theta} \leq C\|f\|_{L^r} \leq C\|g\|_{L_k^r(X;W)}$$

since $\frac{1-\theta}{q} + \frac{\theta}{p} = \frac{1}{r}$. We also have

$$\|h(\theta)\|_{L_k^r(X;W)} \leq C\|\psi^{1/r}\|_{L^r} \leq C\|f\|_{L^r} \leq C\|g\|_{L_p^r(X;W)}.$$

The space $A_\theta \cap L_k^r(X;W)$ has norm

$$\|h(\theta)\|_{A_\theta \cap L_k^r(X;W)} = \|h(\theta)\|_{A_\theta} + \|h(\theta)\|_{L_k^r(X;W)}.$$

We have therefore shown that for every $g \in L_k^r(X;W)$ we can find $h(\theta) \in A_\theta \cap L_k^r(X;W)$ such that

$$\| g-h(\theta) \|_{L_k^r(X;W)} \leq \delta \| g \|_{L_k^r(X;W)}$$

and

$$\| h(\theta) \|_{A_\theta \cap L_k^r(X;W)} \leq C \| g \|_{L_k^r(X;W)} .$$

It follows from Lemma (b) that the inclusion $i: A_\theta \cap L_k^r(X;W) \to L_k^r(X;W)$ is surjective. Therefore $L_k^r(X;W) \subseteq A_\theta$, with a continuous inclusion by the open mapping theorem.

In the other direction, suppose that $h(z) \in \mathcal{H}(A_0, A_1)$. Consider

$$w(z) = \int_X h(z) e^{(z-\theta)^2} W^{\mu(z)}(D)(\varphi \cdot \psi^{1-\lambda(z)}) dX$$

where φ and ψ are smooth with compact support. Then $z \to h(z)$ is continuous and bounded on the strip and holomorphic in the interior, as a map into $A_0 + A_1$ and hence even more as a map into the temperate distributions $\mathcal{S}(X)^*$. Since $z \to e^{(z-\theta)^2} W^{\mu(z)}(D)(\varphi \cdot \psi^{1-\lambda(z)})$ is continuous and bounded on the strip, and holomorphic in the interior as a map into $\mathcal{S}(X)$, it follows that $w(z)$ is continuous and bounded and holomorphic in the interior as an ordinary function.

Using Parseval's relation we can transfer the operator $W^{\mu(z)}(D)$ to $h(z)$ to get

$$w(z) = \int_X \{\overline{W^{\mu(z)}(D)} h(z)\} e^{(z-\theta)^2} \varphi \cdot \psi^{1-\lambda(z)} dx.$$

Let $\frac{1}{p} + \frac{1}{p'} = \frac{1}{q} + \frac{1}{q'} = \frac{1}{r} + \frac{1}{r'} = 1$. Then if $\|\varphi\|_{L^\infty} \leq 1$ and $\|\psi\|_{L^1} \leq 1$ we have

$$\| \varphi \cdot \psi^{1-\lambda(z)} \|_{L^{q'}} \leq C \qquad \text{when Re } z = 0$$

$$\| \varphi \cdot \psi^{1-\lambda(z)} \|_{L^{r'}} \leq C \qquad \text{when Re } z = \theta$$

$$\| \varphi \cdot \psi^{1-\lambda(z)} \|_{L^{p'}} \leq C \qquad \text{when Re } z = 1.$$

Also by Lemma (c)

$$\| \overline{W^{\mu(z)}(D)h(z)} \|_{L^q} \le C(1+|y|)^n \| h(z) \|_{L^q_m(X;W)} \quad \text{when} \quad \text{Re } z = 0$$

and

$$\| \overline{W^{\mu(z)}(D)h(z)} \|_{L^p} \le C(1+|y|)^n \| h(z) \|_{L^p_n(X;W)} \quad \text{when} \quad \text{Re } z = 1.$$

Therefore

$$|w(z)| \le C \|h\|_{\mathfrak{H}(A_0,A_1)} \qquad \text{when} \quad \text{Re } z = 0$$

$$|w(z)| \le C \|h\|_{\mathfrak{H}(A_0,A_1)} \qquad \text{when} \quad \text{Re } z = 1.$$

Applying the three lines theorem to w,

$$|w(\theta)| \le C \|h\|_{\mathfrak{H}(A_0,A_1)}$$

since $\frac{1-\theta}{q'} + \frac{\theta}{p'} = \frac{1}{r'}$. Now choose f so that

$$\int_X \{W^k(D)h(\theta)\} \cdot f \, dX \ge \tfrac{2}{3} \| h(\theta) \|_{L^r_k(X;W)}$$

and $\|f\|_{L^{r'}} \le 1$. Choose φ and ψ to be smooth with compact support so that $\|\varphi\|_{L^\infty} \le 1$, $\|\psi\|_{L^1} \le 1$ and $\| f - \varphi \cdot \psi^{1/r'} \|_{L^{r'}} \le 1/3$.

Then

$$w(\theta) = \int_X \{W^k(D)h(\theta)\} \varphi \, \psi^{1/r'} \, dX \ge 1/3 \ \| h(\theta) \|_{L^r_k(X;W)} .$$

Thus

$$\| h(\theta) \|_{L^r_k(X;W)} \le C \|h\|_{\mathfrak{H}(A_0,A_1)} .$$

Thus

$$\| h(\theta) \|_{L^r_k(X;W)} \le C \| h(\theta) \|_{A_\theta} ,$$

so $A_\theta \subseteq L^r_k(X;W)$ with a continuous inclusion. This shows $A_\theta = L^r_k(X;W)$.

6. Closely related to the potential spaces $L^p_n = L^p_n(X;W_\Sigma)$ are the Besov spaces Λ^p_s which we define for positive non-integral s. As before we assign weight σ_i to the variable x_i and let σ be their least common multiple. Let T^v_j be the operator defined by translation by v in the $j^{\underline{th}}$ coordinate

$$T_j^v f(x_1,\dots,x_j,\dots,x_n) = f(x_1,\dots,x_j + v,\dots,x_n).$$

Then $\Delta_j^v = T_j^v - I$ is the operator defined by differencing by v in the $j^{\underline{th}}$ direction. For $0 < \alpha < 1$ we define the Besov space Λ_α^p to be the space of all f with $\|f\|_{L^p} \leq C$ and for all j

$$\|\Delta_j^v f\|_{L^p} \leq C|v|^{\alpha\sigma_j/\sigma}.$$

We take $\|f\|_{\Lambda_\alpha^p}$ to be the least such constant C. More generally, if s is not an integer we define Λ_s^p to be the space of all those f such that if $\|\gamma\| < s$ then $\|D^\gamma f\|_{L^p} \leq C$ and if $s - \sigma/\sigma_j < \|\gamma\| < s$ then

$$\|\Delta_j^v D^\gamma f\|_{L^p} \leq C|v|^{(s-\|\gamma\|)\sigma_j/\sigma}.$$

We take $\|f\|_{\Lambda_s^p}$ to be the smallest such constant. The advantage of the Besov spaces is the directness of their definition, which is independent of the Fourier transform and hence more suited to non-linear calculations. Their disadvantage is that they are not (so easily) defined for integer or negative s. Our procedure in this paper is to make linear calculations with the L_n^p and non-linear calculations with the Λ_s^p, and pass back and forth by the following theorem, which says they are almost the same, the Besov spaces being just a little larger. Note that our Λ_α^p is Stein's $\Lambda_\alpha^{p,\infty}$.

<u>Theorem</u>. If $r < s$ then

$$L_s^p \subseteq \Lambda_s^p \subseteq L_r^p .$$

<u>Proof</u>. We follow Peetre [18], Chapter X. First we show $L_s^p \subset \Lambda_s^p$. Suppose $s - \sigma/\sigma_j < \|\gamma\| < s$. The fundamental theorem of calculus says

$$\Delta_j^v f(x_1,\dots,x_j,\dots,x_n) = \int_0^v D_j f(x_1,\dots,x_j + u,\dots,x_n)du.$$

Therefore

$$\| \Delta_j^v D^\gamma f\|_{L^p} \leq C|v| \; \| D_j D^\gamma f\|_{L^p}$$

$$\| \Delta_j^v D^\gamma f\|_{L^p} \leq C|v| \; \| f\|_{L^p_{\|\gamma\| +\sigma/\sigma_j}} .$$

But also

$$\| \Delta_j^v \, D^\gamma f\|_{L^p} \leq C\| D^\gamma f\|_{L^p} \leq C \; \| f\|_{L^p_{\|\gamma\|}} .$$

Then we can interpolate $\Delta_j^v D^\gamma$ as a map of $L^p_{\|\gamma\|} \to L^p$ and $L^p_{\|\gamma\| +\sigma/\sigma_j} \to L^p$. We conclude that $\Delta_j^v D^\gamma$ maps L^p_s into L^p and

$$\| \Delta_j^v D^\gamma f\|_{L^p} \leq C|v|^{(s-\|\gamma\|)\sigma_j/\sigma} \; \| f\|_{L^p_s} .$$

Hence $L^p_s \subseteq \Lambda^p_s$.

In the other direction we show $\Lambda^p_s \subseteq L^p_r$ if $r < s$. Introduce the partial smoothing operator S_j^v defined by

$$S_j^v f(x_1,\dots,x_j,\dots,x_n) = \frac{1}{v} \int_0^v f(x_1,\dots,x_j+u,\dots,x_n)du.$$

Then we have the formulas

$$S_j^v f-f = \frac{1}{v} \int_0^v \Delta_j^u f \; du$$

$$D_j S_j^v f = \frac{1}{v} \Delta_j^v f.$$

Suppose $s - \sigma/\sigma_j < \|\gamma\| < s$. Applying these formulas to $D^\gamma f$ we have

$$\| S_j^v D^\gamma f - D^\gamma f \|_{L^p} \leq \frac{1}{v} \int_0^v \| \Delta_j^u D^\gamma f \|_{L^p} \, du$$

$$\leq C|v|^{(s - \|\gamma\|)\sigma_j/\sigma} \, \| f \|_{\Lambda_s^p}$$

$$\| D_j S_j^v D^\gamma f \|_{L^p} \leq \frac{1}{v} \| \Delta_j^v D^\gamma f \|_{L^p}$$

$$\leq C|v|^{-1+(\sigma - \|\gamma\|)\sigma_j/\sigma} \, \| f \|_{\Lambda_s^p} .$$

Let W_j be the partial weight function corresponding to the multiplier $(1 + \xi_j^{2\sigma_j})^{1/2\sigma}$. Then $\|f\|_{L^p_{\sigma/\sigma_j}(W_j)} \leq C(\|f\|_{L^p} + \|D_j f\|_{L^p})$. Applying this to $S_j^v D^\gamma f$ we have

$$\| S_j^v D^\gamma f \|_{L^p} \leq C \| D^\gamma f \|_{L^p} \leq C \| f \|_{\Lambda_s^p}$$

so

$$\| S_j^v D^\gamma f \|_{L^p_{\sigma/\sigma_j}(W_j)} \leq C|v|^{-1+(\sigma - \|\gamma\|)\sigma_j/\sigma} \, \| f \|_{\Lambda_s^p}$$

at least for $|v| \leq 1$. We conclude that

$$\| [S_j^v - S_j^{v/2}] \, D^\gamma f \|_{L^p_{\sigma/\sigma_j}(W_j)} \leq C|v|^{-1+\alpha} \| f \|_{\Lambda_s^p}$$

$$\| [S_j^v - S_j^{v/2}] \, D^\gamma f \|_{L^p_0(W_j)} \leq C|v|^{\alpha} \, \| f \|_{\Lambda_s^p}$$

where we let $\theta = \sigma - \|\gamma\|$ and $|v| \leq 1$. By interpolation it follows that

$$\| [S_j^v - S_j^{v/2}] D^\gamma f \|_{L^p_\theta (W_j)} \leq C|v|^{\alpha - \theta} \, \| f \|_{\Lambda_s^p}$$

for $0 < \theta < 1$. We choose $\theta = (r - \|\gamma\|)\sigma_j/\sigma$. Then provided that r is still close enough to s, $\theta > 0$ and $\theta < \alpha$. Therefore

$$\sum_{k=0}^{\infty} [S_j^{1/2^k} - S_j^{1/2^{k+1}}]D^\gamma f$$

converges absolutely in $L^p_\theta(W_j)$. But we can write

$$[S_j^{1/2^k} - S_j^{1/2^{k+1}}]D^\gamma f = [S_j^{1/2^k} - I]D^\gamma f - [S_j^{1/2^{k+1}} - I]D^\gamma f$$

so that the series telescopes, and $[S_j^v - I]f \to 0$ in L^p as $v \to 0$. Hence the series sums to $[S_j^1 - I]D^\gamma f$ in L^p, and hence in $L^p_\theta(W_j)$ also. Thus

$$\| [S_j^1 - I]D^\gamma f\|_{L^p_\theta(W_j)} \leq \sum_{k=0}^{\infty} C|v|^{(\alpha-\theta)/2^k} \|f\|_{\Lambda^p_s}$$

$$\leq C\|f\|_{\Lambda^p_s}.$$

Also

$$\| S_j^1 D^\gamma f\|_{L^p_\theta(W_j)} \leq C\| S_j^1 D^\gamma f\|_{L^p_{\sigma/\sigma_j}(W_j)}$$

$$\leq C(\| S_j^1 D^\gamma f\|_{L^p} + \| D_j S_j^1 D^\gamma f\|_{L^p}) \leq C\|f\|_{\Lambda^p_s}.$$

Thus

$$\| D^\gamma f\|_{L^p_\theta(W_j)} \leq C\|f\|_{\Lambda^p_s}.$$

Choose $\gamma = (0,\ldots, k_j,\ldots,0)$ with the only non-zero entry k_j in the $j^{\underline{th}}$ place, where k_j is the largest integer less than $s\sigma_j/\sigma$. Then $\|f\|_{L^p_r(W_j)} \leq C\|f\|_{\Lambda^p_s}$. Since this is true for all j, and since W is equivalent to $W_1 + \ldots + W_n$, it follows that $\|f\|_{L^p_r} \leq C\|f\|_{\Lambda^p_s}$, and $\Lambda^p_s \subseteq L^p_r$.

7. Next we show how to use the Besov spaces for non-linear estimates.

Multiplication Theorem. Suppose $0 < \alpha < 1$, $p' \leq p$, $q' \leq q$, $r' \leq r$ and

$$\frac{1}{p} + \frac{1}{q} = \frac{1}{r} \qquad \frac{1}{p'} + \frac{1}{q} = \frac{1}{p} + \frac{1}{q'} = \frac{1}{r'} .$$

If $f \in L^p \cap \Lambda_\alpha^{p'}$ and $g \in L^q \cap \Lambda_\alpha^{q'}$ then the product $fg \in L^r \cap \Lambda_\alpha^{r'}$ and

$$\| fg\|_{\Lambda_\alpha^{r'}} \leq C(\| f\|_{\Lambda_\alpha^{p'}} \| g\|_{L^q} + \| f\|_{L^q} \| g\|_{\Lambda_\alpha^{q'}})$$

$$\| fg\|_{L^r} \leq C\| f\|_{L^p} \| g\|_{L^q} .$$

Proof. Clearly $fg \in L^r$. We must estimate the difference $\Delta_j^v(fg)$. Now

$$\Delta_j^v(fg) = \Delta_j^v f \cdot T_j^v g + f \cdot \Delta_j^v g .$$

Since L^p norms are invariant under the translation T_j^v

$$\| \Delta_j^v(fg)\|_{L^{r'}} \leq \| \Delta_j^v f\|_{L^{p'}} \| T_j^v g\|_{L^q} + \| f\|_{L^p} \| \Delta_j^v g\|_{L^{q'}}$$

$$\| fg\|_{\Lambda_\alpha^{r'}} \leq C(\| f\|_{\Lambda_\alpha^{p'}} \| g\|_{L^q} + \| f\|_{L^p} \| g\|_{\Lambda_\alpha^{q'}}).$$

Compositon Theorem. Let $\Gamma(x,z)$ be a continuous non-linear function defined on a set Ω and vanishing for x outside a compact set. Suppose Γ satisfies a Lipschitz condition

$$|\Gamma(x,w) - \Gamma(x,z)| \leq C|w-z| .$$

Define Γf for any function f whose graph lies in Ω by

$$\Gamma f(x) = \Gamma(x,f(x)).$$

Then $\| \Gamma f\|_{\Lambda^p_\alpha} \leq C(\| f\|_{\Lambda^p_\alpha} + 1)$.

Proof. $|\Gamma(x,z)| \leq C(1+|z|)$ from the Lipschitz condition. Thus

$$\| \Gamma f\|_{L^p} \leq C(\| f\|_{L^p} + 1).$$

Also

$$|\Delta^j_v \Gamma(f)| \leq C|\Delta^j_v f|$$

so

$$\| \Delta^j_v \Gamma(f)\|_{L^p} \leq C\| \Delta^j_v f\|_{L^p}.$$

Thus

$$\| \Gamma(f)\|_{\Lambda^p_\alpha} \leq C(\| f\|_{\Lambda^p_\alpha} + 1).$$

We remark that the same holds if f is a system of functions $f = \{f^1,\ldots,f^N\}$ with $\| f\|_{\Lambda^p_\alpha} = \Sigma\| f^i\|_{\Lambda^p_\alpha}$, and $\Gamma(f)$ is also a system.

8. In order to study functions on a half space we use an extension operator similar to that in Stein [22]. Its construction depends on the following Lemma.

Lemma. Let

$$\varphi(x) = \frac{e^4}{\pi} \cdot \frac{e^{-(x^{1/4}+x^{-1/4})}\sin(x^{1/4}-x^{-1/4})}{1+x} .$$

Then for every positive or negative integer n

$$\int_0^\infty x^n|\varphi(x)|dx < \infty$$

$$\int_0^\infty x^n\varphi(x)dx = (-1)^n.$$

Also $\varphi(\frac{1}{x}) = -x\varphi(x)$.

Proof. The absolute convergence of the integral is immediate.

Let $z^{1/4}$ be the complex-analytic function on the plane slit along the positive x-axis which equals $x^{1/4}$ on the top slit. Then $z^{1/4}$ equals $i\,x^{1/4}$ on the bottom slit. Let

$$f(z) = \frac{e^{-(1-i)z^{1/4} - (1+i)z^{-1/4}}}{1+z}.$$

Then

$$f(x) = \frac{e^{-(x^{1/4}+x^{-1/4})}\{\cos(x^{1/4}-x^{-1/4}) + i\,\sin(x^{1/4}-x^{-1/4})\}}{1+x}$$

on the top slit and the conjugate on the bottom slit. Let γ be the path coming in on the bottom slit and going out on the top. Then

$$\int_\gamma z^n f(z)\,dz = 2i\int_0^\infty \frac{x^n\, e^{-(x^{1/4}+x^{-1/4})}\sin(x^{1/4}-x^{-1/4})}{1+x}\,dx.$$

On the other hand $f(z)$ decreases very rapidly at ∞, so the integral can be evaluated by the residue at $z=-1$:

$$\int_\gamma z^n f(z)\,dz = 2\pi i\cdot(-1)^n\cdot e^{-4}$$

since $(-1)^{1/4} = 1+i$ for our choice of $z^{1/4}$. Thus

$$\int_0^\infty \frac{x^n e^{-(x^{1/4}+x^{-1/4})}\sin(x^{1/4}-x^{-1/4})}{1+x}\,dx = \frac{\pi}{e^4}(-1)^n .$$

The result follows. The relation $\varphi(1/x) = -x\varphi(x)$ follows from a direct verification.

9. In order to study functions on a half space we distinguish the last variable. Let $X \times Y$ be the space of the variables $\{x_1,\ldots,x_n,y\}$. Let $Y^+ = \{y \geq 0\}$ and $Y^- = \{y \leq 0\}$. We denote by $\mathcal{S}(X\times Y^+)$ the smooth rapidly decreasing functions on the half-space $X \times Y^+$, and we denote by $\mathcal{S}(X\times Y^+/0)$ the subspace of those functions which vanish on the boundary $X\times\{0\}$ together

with all of their derivatives. Similarly for $\mathcal{S}(X \times Y^-)$ and $\mathcal{S}(X \times Y^-/0)$ on the other side. There is a natural short exact sequence

$$0 \to \mathcal{S}(X \times Y^-/0) \xrightarrow{Z_-} \mathcal{S}(X \times Y) \xrightarrow{C_+} \mathcal{S}(X \times Y^+) \to 0$$

where Z_- is the "zero extension" and C_+ is the "cutoff". We define the extension

$$E_+ : \mathcal{S}(X \times Y^+) \to \mathcal{S}(X \times Y)$$

by the formula

$$E_+f(x,y) = f(x,y) \quad \text{for} \quad y \geq 0$$

$$E_+f(x,-y) = \int_0^\infty \varphi(\lambda) f(x,\lambda y)\, d\lambda$$

where $\varphi(\lambda)$ is chosen as before.

<u>Theorem.</u> If f is smooth then E_+f is smooth and $E_+ : \mathcal{S}(X \times Y^+) \to \mathcal{S}(X \times Y)$ is a continuous linear map.

<u>Proof.</u> For $y \geq 0$

$$D_x^\alpha \, D_y^n \, E_+f(x,-y) = (-1)^n \int_0^\infty \lambda^n \varphi(\lambda) f(x,\lambda y)\, d\lambda \; .$$

Then as $y \to 0$

$$D_x^\alpha \, D_y^n \, E_+f(x,-y) \to D_x^\alpha \, D_y^n \, f(x,0)$$

since $\int_0^\infty \lambda^n \varphi(\lambda)\, d\lambda = (-1)^n$. Finally

$$\sup |D_x^\alpha \, D_y^n \, E_+f| \leq C \sup |D_x^\alpha \, D_y^n \, f|$$

where $C = \int_0^\infty \lambda^n |\varphi(\lambda)|\, d\lambda$.

There is a similar extension E_- on the other half space. Clearly $C_+E_+ = I$ and $C_-E_- = I$. Therefore $(E_+C_+)^2 = E_+C_+$ so E_+C_+ is a continuous projection. If $f \in \mathcal{S}(X \times Y)$ then E_+C_+f

equals f on $X \times Y^+$, so $(I-E_+C_+)f$ equals zero on $X \times Y^+$. Therefore there is a unique element R_-f in $\mathcal{S}(X \times Y^-/0)$ with $Z_-R_-f = (I-E_+C_+)f$. This defines a continuous retraction map

$$R_-: \mathcal{S}(X \times Y) \to \mathcal{S}(X \times Y^-/0).$$

Then Z_-R_- is again a projection and $R_-Z_- = I$. This proves the following:

<u>Theorem</u>. The operators E_+ and R_- split the sequence for Z_- and C_+

$$0 \underset{\longleftarrow}{\longrightarrow} \mathcal{S}(X \times Y^-/0) \underset{R_-}{\overset{Z_-}{\rightleftarrows}} \mathcal{S}(X \times Y) \underset{E_+}{\overset{C_+}{\rightleftarrows}} \mathcal{S}(X \times Y^+) \underset{\longleftarrow}{\longrightarrow} 0$$

in the sense that

$$C_+ Z_- = 0 \qquad R_- E_+ = 0$$

$$R_- Z_- = 0 \qquad C_+ E_+ = I$$

$$E_+ C_+ + Z_- R_- = I.$$

There is a direct sum decomposition

$$\mathcal{S}(X \times Y) \approx \mathcal{S}(X \times Y^-/0) \oplus \mathcal{S}(X \times Y^+)$$

given by the projections $E_+ C_+$ and Z_-R_-.

There is a natural pairing

$$\mathcal{S}(X \times Y^+/0) \times \mathcal{S}(X \times Y^+) \to \mathbb{C}$$

such that derivatives D^α are self-adjoint:

$$\langle D^\alpha f, g\rangle = \langle f, D^\alpha g\rangle.$$

This gives natural continuous dense inclusions in the dual spaces

$$\mathcal{S}(X \times Y^+) \to \mathcal{S}(X \times Y^+/0)^* \quad \text{and} \quad \mathcal{S}(X \times Y^+/0) \to \mathcal{S}(X \times Y^+)^*.$$

<u>Theorem</u>. If $f \in \mathcal{S}(X \times Y^+)$ and $g \in \mathcal{S}(X \times Y)$

$$\langle E_+ f, g\rangle = \langle f, R_+ g\rangle.$$

Proof. This follows since $\varphi(1/\lambda) = -\lambda\varphi(\lambda)$.

$$\langle E_+f, g\rangle = \int_X \int_0^\infty \int_0^\infty \varphi(\lambda) f(x,\lambda y) g(-y)\, dy\, d\lambda$$

$$+ \int_X \int_0^\infty f(x,y) g(y)\, dy.$$

Substitute $\mu = 1/\lambda$ and $z = \lambda y$ in the first integral and $z = y$ in the second.

$$\langle E_+f, g\rangle = \int_X \int_0^\infty f(x,z) g(z)\, dz$$

$$- \int_X \int_0^\infty \int_0^\infty \varphi(\mu) f(x,z) g(-\mu z)\, d\mu\, dy$$

$$= \langle f, R_+g\rangle.$$

It follows that the extension E_+ also extends distributions to distributions by duality, in the sense that the following diagram commutes:

$$\begin{array}{ccc} \mathcal{S}(X \times Y^+) & \xrightarrow{E_+} & \mathcal{S}(X \times Y) \\ \downarrow & & \downarrow \\ \mathcal{S}(X \times Y^+/0)^* & \xrightarrow{R_+^*} & \mathcal{S}(X \times Y)^* . \end{array}$$

In fact the direct sum splitting of $\mathcal{S}(X \times Y)$ also extends to distributions, in the sense that we have a commutative diagram

$$\begin{array}{ccccccccc} 0 & \rightleftarrows & \mathcal{S}(X \times Y^-/0) & \underset{R_-}{\overset{Z_-}{\rightleftarrows}} & \mathcal{S}(X \times Y) & \underset{E_+}{\overset{C_+}{\rightleftarrows}} & \mathcal{S}(X \times Y^+) & \rightleftarrows & 0 \\ & & \downarrow & & \downarrow & & \downarrow & & \\ 0 & \rightleftarrows & \mathcal{S}(X \times Y^-)^* & \underset{E_-^*}{\overset{C_-^*}{\rightleftarrows}} & \mathcal{S}(X \times Y)^* & \underset{R_+^*}{\overset{Z_+^*}{\rightleftarrows}} & \mathcal{S}(X \times Y^+/0)^* & \rightleftarrows & 0. \end{array}$$

10. We can now define spaces L_n^p on $X \times Y^+$. Let $x_1,\ldots,x_n$ have weights $\sigma_1,\ldots,\sigma_n$ and let y have weight ρ, with $\sigma = \ell.c.m.\{\sigma_1,\ldots,\sigma_n,\rho\}$. If $\xi_1,\ldots,\xi_n$ and η denote the dual variables then $L_n^p(X\times Y) = \{f \in \mathcal{S}(X\times Y)^* : W^n(D_x,D_y)f \in L^p(X\times Y)\}$ where $W(\xi,\eta) = (1+\xi_1^{2\sigma_1}+\ldots+\xi_n^{2\sigma_n}+\eta^{2\rho})^{1/2\sigma}$. We define

$$L_k^p(X\times Y^+) = \{f \in \mathcal{S}(X\times Y^+/0)^* : \exists g \in L_k^p(X\times Y) \text{ with } Z_+^* g = f\}$$

with norm

$$\|f\|_{L_k^p(X\times Y^+)} = \inf\{\|g\|_{L_k^p(X\times Y)} : Z_+^* g = f\}.$$

Also we define

$$L_k^p(X\times Y^-/0) = \{f \in \mathcal{S}(X\times Y^-)^* : C_-^* f \in L_k^p(X\times Y)\}$$

with norm

$$\|f\|_{L_k^p(X\times Y^-/0)} = \|C_-^* f\|_{L_k^p(X\times Y)} .$$

The spaces $L_k^p(X\times Y^-)$ and $L_k^p(X\times Y^+/0)$ are defined similarly.

<u>Lemma</u>. The projections $C_-^*E_-^*$ and $R_+^*Z_+^*$ on $\mathcal{S}(X\times Y)^*$ induce bounded linear projections on $L_k^p(X\times Y)$ for $1 < p < \infty$ and all k.

<u>Proof</u>. First suppose k is a positive integer multiple of σ. Then $f \in L_k^p(X\times Y)$ if and only if $D_x^\alpha D_y^\beta f \in L^p(X\times Y)$ whenever

$$\|(\alpha,\beta)\| = \alpha_1\sigma_1+\ldots+\alpha_n\sigma_n+\beta\rho \leq k$$

and

$$\|f\|_{L_k^p(X\times Y)} \leq \sum_{\|(\alpha,\beta)\|\leq k} \|D_x^\alpha D_y^\beta f\|_{L^p(X\times Y)} .$$

On the dense subspace $\mathcal{S}(X\times Y)$ the operator $R_+^*Z_+^*$ is given by E_+C_+, and for $y \geq 0$

$$E_+C_+f(x,y) = f(x,y)$$

$$E_+C_+f(x,-y) = \int_0^\infty \varphi(\lambda)f(x,\lambda y)d\lambda.$$

Thus

$$D_x^\alpha D_y^\beta E_+C_+f(x,-y) = (-1)^\beta \int_0^\infty \lambda^\beta \varphi(\lambda) D_x^\alpha D_y^\beta f(x,\lambda z)d\lambda$$

$$\| D_x^\alpha D_y^\beta E_+C_+f\|_{L^p(X\times Y)} \leq C\| D_x^\alpha D_y^\beta f\|_{L^p(X\times Y)}$$

$$\| E_+C_+f\|_{L_k^p(X\times Y)} \leq C\| f\|_{L_k^p(X\times Y)} .$$

If $-k$ is a negative multiple of σ then $f \in L_{-k}^p(X\times Y)$ if and only if we can find $g_{\alpha\beta} \in L^p(X\times Y)$ with $f = \sum_{\|(\alpha,\beta)\| \leq k} D_x^\alpha D_y^\beta g_{\alpha\beta}$ and

$$\| f\|_{L_{-k}^p(X\times Y)} = \inf\{ \sum_{\|(\alpha,\beta)\| \leq k} \| g_{\alpha\beta}\|_{L^p(X\times Y)} \}.$$

Again if f is in the dense subspace $\mathcal{S}(X \times Y)$ we can choose $g_{\alpha\beta} \in \mathcal{S}(X \times Y)$ also, and

$$E_+C_+f(x,-y) = \Sigma\, E_+C_+D_x^\alpha D_y^\beta g_{\alpha\beta}(x,-y)$$

$$= \Sigma\, D_x^\alpha D_y^\beta \{(-1)^\beta \int_0^\infty \lambda^{-\beta}\varphi(\lambda) g_{\alpha\beta}(x,\lambda y)d\lambda\}$$

and

$$\| \int_0^\infty \lambda^{-\beta}\varphi(\lambda) g_{\alpha\beta}(x,\lambda y)d\lambda\|_{L^p(X\times Y)} \leq C\| g_{\alpha\beta}\|_{L^p(X\times Y)}.$$

Thus $\| E_+C_+f\|_{L_{-k}^p(X\times Y)} \leq C\| f\|_{L_{-k}^p(X\times Y)}$. The result now follows for all real values of k by interpolation. A similar argument applies to the other projection.

We immediately deduce the following result:

Theorem. There is a split-exact sequence

$$0 \rightleftarrows L^p_k(X \times Y^-/0) \underset{E^*_-}{\overset{C^*_-}{\rightleftarrows}} L^p_k(X \times Y) \underset{R^*_+}{\overset{Z^*_+}{\rightleftarrows}} L^p_k(X \times Y^+) \rightleftarrows 0$$

giving rise to a direct sum decomposition

$$L^p_k(X \times Y) \approx L^p_k(X \times Y^-/0) \oplus L^p_k(X \times Y^+)$$

for $1 < p < \infty$ and all k.

Thus E_+ defines a bounded linear extension in all the norms L^p_k. Also we have explicit characterizations when k is an integer multiple of σ.

Theorem. If k is a positive integer multiple of σ then if $f \in \mathcal{S}(X \times Y^+/0)^*$

$$f \in L^p_k(X \times Y^+) \Longleftrightarrow D^\alpha_x D^\beta_y f \in L^p(X \times Y^+) \quad \text{for} \quad \|(\alpha,\beta)\| \le k$$

$$f \in L^p_{-k}(X \times Y^+) \Longleftrightarrow \exists g_{\alpha\beta} \in L^p(X \times Y^+) \quad \text{with}$$

$$f = \sum_{\|(\alpha,\beta)\| \le k} D^\alpha_x D^\beta_y g_{\alpha\beta}$$

and if $f \in \mathcal{S}(X \times Y^+)^*$ then

$$f \in L^p_k(X \times Y^+/0) \Longleftrightarrow D^\alpha_x D^\beta_y f \in L^p_k(X \times Y^+) \quad \text{for} \quad \|(\alpha,\beta)\| \le k$$

$$f \in L^p_{-k}(X \times Y^+/0) \Longleftrightarrow \exists g_{\alpha\beta} \in L^p(X \times Y^+) \quad \text{with}$$

$$f = \sum_{\|(\alpha,\beta)\| \le k} D^\alpha_x D^\beta_y g_{\alpha\beta} .$$

Thus the difference between $L^p_k(X \times Y^+)$ and $L^p_k(X \times Y^+/0)$ depends only on the kind of distribution which we consider, i.e., $\mathcal{S}(X \times Y^+)^*$ or $\mathcal{S}(X \times Y^+/0)^*$.

The spaces $L^p(X \times Y)$ and $L^q(X \times Y)$ are dual if $\frac{1}{p} + \frac{1}{q} = 1$. By Parseval's relation it follows that $L^p_k(X \times Y)$ and $L^q_{-k}(X \times Y)$ are also dual. Then from the direct sum decomposition we have that $L^p_k(X \times Y^+)$ and $L^q_{-k}(X \times Y^+/0)$ are dual, as are $L^p_{-k}(X \times Y^+)$ and $L^q_k(X \times Y^+/0)$. Hence in particular all these spaces are reflexive Banach spaces. Finally, we remark that using the direct sum decomposition, if we interpolate between two spaces $L^p_k(X \times Y^+)$ or two spaces $L^p_k(X \times Y^+/0)$ we obtain a third of the same type, according to the rule given in the Interpolation Theorem.

11. We consider more closely the relation between $L^p_k(X \times Y^+)$ and $L^p_k(X \times Y^+/0)$. There is a natural map for all k

$$Z^*_+ \; C^*_+ \colon L^p_k(X \times Y^+/0) \to L^p_k(X \times Y^+).$$

<u>Theorem</u>. If $1 < p < \infty$ then $L^p_k(X \times Y^+/0) \to L^p_k(X \times Y^+)$ is an isomorphism for $-1 + 1/p < \rho k/\sigma < 1/p$.

<u>Proof</u>. It is sufficient to prove the result for $0 \leq \rho k/\sigma < 1/p$ since the other case follows from this by duality. The norm in $L^p_k(X \times Y)$ is defined by a weight function $W(\xi, \eta) = (1 + \xi_1^{2\sigma_1} + \dots + \xi_n^{2\sigma_n} + \eta^{2\rho})^{1/2\sigma}$. Define partial weight functions $W(\xi) = (1 + \xi_1^{2\sigma_1} + \dots + \xi_n^{2\sigma_n})^{1/2\sigma}$ and $W(\eta) = (1 + \eta^{2\rho})^{1/2\sigma}$. Then $W(\xi, \eta)^k$ is comparable to $W(\xi)^k + W(\eta)^k$. In particular if $k = \sigma/\rho$ we have

<u>Lemma (a)</u>. A function $f \in L^p_{\sigma/\rho}(X \times Y)$ if and only if $W(D_x)^{\sigma/\rho} f \in L^p(X \times Y)$ and $D_y f \in L^p(X \times Y)$; moreoever the norms

$$\| f \|_{L^p_{\sigma/\rho}(X \times Y)} \quad \text{and} \quad \| W(D_x)^{\sigma/\rho} f \|_{L^p(X \times Y)} + \| D_y f \|_{L^p(X \times Y)}$$

are equivalent.

<u>Proof</u>. Let $E_0(\eta) = 1/(1+\eta^{2\rho})^{1/2\rho}$ and $E_1(\eta) = \eta^{2\rho-1}/(1+\eta^{2\rho})^{(2\rho-1)/2\rho}$. Then $W^{\sigma/\rho}(\eta) = (1+\eta^{2\rho})^{1/2\rho}$

$= E_0(\eta) + E_1(\eta)\cdot\eta$. Let $E(\xi,\eta) = W^{\sigma/\rho}(\xi,\eta)/[W^{\sigma/\rho}(\xi)+W^{\sigma/\rho}(\eta)]$. Then $E_0(D_y), E_1(D_y)$ and $E(D_x, D_y)$ define bounded linear maps of $L^p(X\times Y)$ into itself and

$$W^{\sigma/\rho}(D_x, D_y) = E(D_x, D_y)\{W^{\sigma/\rho}(D_x) + E_0(D_y) + E_1(D_y)\cdot D_y\}.$$

This proves the lemma one way. For the converse let $F_0(\xi,\eta) = W^{\sigma/\rho}(\xi)/W^{\sigma/\rho}(\xi,\eta)$ and $F_1(\xi,\eta) = \eta/W^{\sigma/\rho}(\xi,\eta)$. Then $F_0(D_x, D_y)$ and $F_1(D_x, D_y)$ are bounded linear operators on $L^p(X\times Y)$ and

$$W^{\sigma/\rho}(D_x) = F_0(D_x, D_y)\; W^{\sigma/\rho}(D_x, D_y)$$

$$D_y = F_1(D_x, D_y)\; W^{\sigma/\rho}(D_x, D_y).$$

This proves the lemma the other way also.

Next we note that the operator $W(D_x)^{\sigma/\rho}$ by adjoints takes $\mathcal{S}(X\times Y)^*$ into itself and preserves $\mathcal{S}(X\times Y^-)^*$, so it takes the subspace $L^p_{\sigma/\rho}(X\times Y^-/0)$ into $L^p(X\times Y^-)$ and hence acts naturally also on $\mathcal{S}(X\times Y^+/0)^*$ and defines a map of $L^p_{\sigma/\rho}(X\times Y^+)$ into $L^p(X\times Y^+)$, as can be seen from the exact sequences

$$\begin{array}{ccccccccc}
0 \to & L^p_{\sigma/\rho}(X\times Y^-/0) & \xrightarrow{Z_-} & L^p_{\sigma/\rho}(X\times Y) & \xrightarrow{C_+} & L^p_{\sigma/\rho}(X\times Y^+) & \to 0 \\
& \downarrow W(D_x)^{\sigma/\rho} & & \downarrow W(D_x)^{\sigma/\rho} & & \downarrow W(D_x)^{\sigma/\rho} & \\
0 \to & L^p(X\times Y^-) & \xrightarrow{Z_-} & L^p(X\times Y) & \xrightarrow{C_+} & L^p(X\times Y^+) & \to 0
\end{array}$$

Also D_x commutes with the extension E_+; and so $W^{\sigma/\rho}(D_x)$ also commutes with E_+ in the diagram

$$\begin{array}{ccc}
\mathcal{S}(X\times Y^+) & \xrightarrow{E_+} & \mathcal{S}(X\times Y) \\
\downarrow W^{\sigma/\rho}(D_x) & & \downarrow W^{\sigma/\rho}(D_x) \\
\mathcal{S}(X\times Y^+) & \xrightarrow{E_+} & \mathcal{S}(X\times Y)
\end{array}$$

as can be seen by considering the partial Fourier transform in the variables $(x_1,\dots,x_n)$ which is sufficient to apply $W^{\sigma/\rho}(D_x)$. Thus there is a commutative diagram

$$\begin{array}{ccc} L^p_{\sigma/\rho}(X\times Y^+) & \xrightarrow{R_+^*} & L^p_{\sigma/\rho}(X\times Y) \\ \Big\downarrow W^{\sigma/\rho}(D_x) & & \Big\downarrow W^{\sigma/\rho}(D_x) \\ L^p(X\times Y^+) & \xrightarrow{R_+^*} & L^p(X\times Y) \end{array}$$

Lemma (b). A function $f \in L^p_{\sigma/\rho}(X\times Y^+)$ if and only if $f \in \mathcal{S}(X\times Y^+/0)^*$, $W^{\sigma/\rho}(D_x)f \in L^p(X\times Y^+)$ and $D_y f \in L^p(X\times Y^+)$.

Proof. $f \in L^p_{\sigma/p}(X\times Y^+) \Leftrightarrow E_+f \in L^p_{\sigma/p}(X\times Y) \Leftrightarrow W^{\sigma/\rho}(D_x)E_+f \in L^p(X\times Y)$ and $D_yE_+f \in L^p(X\times Y)$. But $W^{\sigma/\rho}(D_x)E_+f = E_+W^{\sigma/\rho}(D_x)f$ and

$$D_y\, E_+f(x,-y) = -\int_0^\infty \lambda\varphi(\lambda)D_yf(x,\lambda y)\,d\lambda$$

so

$$D_y\, E_+f \in L^p(X\times Y) \Leftrightarrow D_yf \in L^p(X\times Y^+).$$

We also clearly have $\|f\|_{L^p_{\sigma/\rho}(X\times Y^+)}$ equivalent to

$$\|W^{\sigma/\rho}(D_x)f\|_{L^p(X\times Y^+)} + \|D_yf\|_{L^p(X\times Y^+)}$$

Now $Z_+^* \oplus Z_-^*: L^p(X\times Y) \to L^p(X\times Y^+) \oplus L^p(X\times Y^-)$ is an isomorphism. Therefore we have

Lemma (c).

$$\|f\|_{L^p_{\sigma/\rho}(X\times Y)} \leq C(\|Z_+^*f\|_{L^p_{\sigma/\rho}(X\times Y^+)} + \|Z_-^*f\|_{L^p_{\sigma/\rho}(X\times Y^-)}).$$

Lemma (d). Let $f_+ \in \mathcal{S}(X\times Y^+)$ and $f_- \in \mathcal{S}(X\times Y_-)$. Suppose $f_+ = f_-$ on $X\times\{0\}$. Let f be defined on $X\times Y^+$ by $f = f_+$ on $X\times Y^+$ and $f = f_-$ on $X\times Y^-$. Then $f \in L^p_{\sigma/\rho}(X\times Y)$ and $\|f\|_{L^p_{\sigma/\rho}(X\times Y)} \leq C(\|f_+\|_{L^p_{\sigma/\rho}(X\times Y^+)} + \|f_-\|_{L^p_{\sigma/\rho}(X\times Y^-)})$.

Proof. By the preceding Lemmas it is only necessary to verify that the function $D_y f$ equal to $D_y f_+$ on $X \times Y^+$ and $D_y f_-$ on $X \times Y^-$ is the distributional derivative of f in $L^p(X \times Y)$. This follows from the cancellation of the boundary terms. Thus if $g \in \mathcal{S}(X \times Y)$

$$\langle f, D_y g\rangle = \int_X \int_{-\infty}^{0} f_-(x,y) D_y g(x,y)\,dx\,dy$$

$$+ \int_X \int_0^{\infty} f_+(x,y)\ D_y g(x,y)\,dx\,dy$$

$$= \int_X [f_-(x,0) - f_+(x,0)]\ g(x,0)\,dx$$

$$- \int_X \int_{-\infty}^{0} D_y f_-(x,y) g(x,y)\,dx\,dy$$

$$- \int_X \int_0^{\infty} D_y f_+(x,y) g(x,y)\,dx\,dy$$

$$= - \langle D_y f_-, g\rangle - \langle D_y f_+, g\rangle.$$

For all $\lambda > 0$ define a shrinking operator

$$M_+(\lambda): \mathcal{S}(X \times Y^+) \to \mathcal{S}(X \times Y^+)$$

$$M_+(\lambda) f(x,y) = f(x, \lambda y).$$

Also define a dual operator

$$N_+(\lambda): \mathcal{S}(X \times Y^+/0) \to \mathcal{S}(X \times Y^+/0)$$

$$N_+(\lambda) g(x,y) = \lambda^{-1} g(x, \lambda^{-1} y)$$

Then $\langle M_+(\lambda) f, g\rangle = \langle f, N_+(\lambda) g\rangle$ so there is a commutative diagram

$$\begin{array}{ccc} \mathcal{S}(X \times Y^+) & \xrightarrow{M_+(\lambda)} & \mathcal{S}(X \times Y^+) \\ \downarrow & & \downarrow \\ \mathcal{S}(X \times Y^+/0)^* & \xrightarrow[N_+^*(\lambda)]{} & \mathcal{S}(X \times Y^+/0)^*. \end{array}$$

Lemma (e). For all $k \geq 0$ and $\lambda \geq 1$, $N_+^*(\lambda)$ defines a bounded linear map of $L_k^p(X \times Y^+)$ into itself and

$$\| N_+^*(\lambda) f \|_{L_k^p(X \times Y^+)} \leq C\lambda^{\frac{\rho k}{\sigma} - \frac{1}{p}} \| f \|_{L_k^p(X \times Y^+)}.$$

Proof. When k is an integer multiple of σ this results from a direct calculation. In fact if $f \in \mathcal{S}(X \times Y^+)$

$$D_x^\alpha M_+(\lambda) f = M_+(\lambda) D_x^\alpha f$$

$$D_y^\alpha M_+(\lambda) f = \lambda M_+(\lambda) D_y^\alpha f.$$

For intermediary values of k it follows by interpolation. In particular

$$\| N_+^*(\lambda) f \|_{L_{\sigma/\rho}^p(X \times Y^+)} \leq C\lambda^{1-1/p} \| f \|_{L_{\sigma/\rho}^p(X \times Y^+)}.$$

Next if $f \in \mathcal{S}(X \times Y^+)$ let $E_+ f \in \mathcal{S}(X \times Y)$ be its extension to $X \times Y$, and $C_- E_+ f \in \mathcal{S}(X \times Y^-)$ the restriction of the extension to the other side. Then $M_-(\lambda) C_- E_+ f \in \mathcal{S}(X \times Y^-)$ and agrees with f on $X \times 0$. Define

$$\widetilde{M}(\lambda) f = \begin{cases} f & \text{on } X \times Y^+ \\ M_-(\lambda) C_- E_+ f & \text{on } X \times Y^- \end{cases}.$$

Then by Lemma (d), $\widetilde{M}(\lambda) f \in L_{\sigma/\rho}^p(X \times Y)$ and

$$\| \widetilde{M}(\lambda) f \|_{L_{\sigma/\rho}^p(X \times Y)} \leq C(\| f \|_{L_{\sigma/\rho}^p(X \times Y^+)} + \| M_-(\lambda) C_- E_+ f \|_{L_{\sigma/\rho}^p(X \times Y^-)})$$

so

$$\|\widetilde{M}(\lambda)f\|_{L^p_{\sigma/\rho}(X\times Y)} \leq C\lambda^{1-1/p}\|f\|_{L^p_{\sigma/\rho}(X\times Y^+)}$$

for $\lambda \geq 1$. Moreover

$$\widetilde{M}(2\lambda)f = \widetilde{M}(\lambda)f \quad \text{on} \quad X\times Y^+ \quad \text{so}$$

$$\|\widetilde{M}(2\lambda)f-\widetilde{M}(\lambda)f\|_{L^p_{\sigma/\rho}(X\times Y)} \leq C\lambda^{1-1/p}\|f\|_{L^p_{\sigma/\rho}(X\times Y^+)}$$

for $\lambda \geq 1$. But now we also have

$$\|\widetilde{M}(2\lambda)f-\widetilde{M}(\lambda)f\|_{L^p(X\times Y)} \leq C\lambda^{-1/p}\|f\|_{L^p(X\times Y^+)}.$$

Therefore by interpolation, for $0 \leq k \leq \sigma/\rho$

$$\|\widetilde{M}(2\lambda)f-\widetilde{M}(\lambda)f\|_{L^p_k(X\times Y)} \leq C\lambda^{\rho k/\sigma - 1/p}\|f\|_{L^p_k(X\times Y^+)}$$

for $\lambda \geq 1$. Thus if $0 \leq \rho k/\sigma < 1/p$

$$\sum_{i=0}^{\infty}\|\widetilde{M}(2^{i+1})f-\widetilde{M}(2^i)f\|_{L^p_k(X\times Y)} \leq C\|f\|_{L^p_k(X\times Y^+)}$$

with $C < \infty$ by the convergence of the geometric series. Consequently $\widetilde{M}(2^i)f$ is a Cauchy sequence in $L^p_k(X\times Y)$. Let

$$\widetilde{M}f = \lim_{i\to\infty}\widetilde{M}(2^i)f.$$

Then $\|\widetilde{M}f\|_{L^p_k(X\times Y)} \leq C\|f\|_{L^p_k(X\times Y)}$. Consider the restriction

$Z^*_-: \mathcal{S}(X\times Y) \to \mathcal{S}(X\times Y^-/0)^*$, given by $\langle Z^*_-f,g\rangle = \langle f,Z_-g\rangle$. Then $Z^*_-\widetilde{M}(\lambda)f = M_-(\lambda)C_-E_+f \in \mathcal{S}(X\times Y^-) \subseteq \mathcal{S}(X\times Y^-/0)^*$. Thus

$$Z^*_-\widetilde{M}f = \lim_{i\to\infty} Z^*_-\widetilde{M}(2^i)f = \lim_{i\to\infty} M_-(2^i)\,C_-\,E_+f.$$

But $\|M_-(\lambda)g\|_{L^p(X\times Y^-)} \leq C\lambda^{-1/p}\|g\|_{L^p(X\times Y^-)}$ so $M_-(\lambda)g \to 0$ in $L^p(X\times Y^-)$ as $\lambda\to\infty$. Thus $Z^*_-\widetilde{M}f = 0$. Hence $\widetilde{M}f$ belongs to the image of C^*_+; that is, there exists a unique element $Mf \in \mathcal{S}(X \times Y^+)^*$ with $C_+^* Mf = \tilde{M}f$. Then $Mf \in L_k^p(X \times Y^+/0)$ and $\| Mf \|_{L_k^p(X\times Y^+/0)} = \| \tilde{M}f \|_{L_k^p(X\times Y)} \leq c \| f \|_{L_k^p(X\times Y^+)}$.

Moreover $Z_+^* \, C_+^* \, Mf = Z_+^* \, \tilde{M}f = \lim_{i\to\infty} Z_+^* \, \tilde{M}(2^i)f$; but $Z_+^* \, M(2^i)f = f$ for all i.

We have shown then that for every $f \in \mathcal{S}(X\times Y^+)$ there is an element $Mf \in L_k^p(X\times Y^+/0)$ with $Z_+^* \, C_+^* \, Mf = f$ and

$$\| Mf \|_{L_k^p(X\times Y^+/0)} \leq c \| f \|_{L_k^p(X\times Y^+)} .$$

Since $\mathcal{S}(X\times Y^+)$ is dense in $L_k^p(X\times Y^+)$ it follows immediately that

$$L_k^p(X\times Y^+/0) \approx L_k^p(X \times Y^+)$$

for $1 < p < \infty$ and $0 < \rho k/\sigma < 1/p$.

12. Outside this range the previous theorem fails. In fact we have the following result

<u>Theorem</u>. If $\rho k/\sigma > 1/p$ then the restriction to the boundary

$$B: \mathcal{S}(X \times Y^+) \to \mathcal{S}(X)$$

given by

$$Bf(x) = f(x,0)$$

extends by completion to a continuous linear map

$$B: L_k^p(X \times Y^+) \to L^p(X).$$

<u>Proof</u>. We know that

$$\| D_y f \|_{L^p(X\times Y^+)} \leq c \| f \|_{L^p_{\sigma/\rho}(X\times Y^+)} .$$

Also if $1 < p < \infty$ and $f \in \mathcal{S}(X \times Y^+)$ then $|f(x,y)|^p$ is continuously differentiable. By the fundamental theorem of calculus

$$\| Bf\|_{L^p(x)} = \int_X |f(x,0)|^p dx = \int_X \int_{y=0}^{\infty} \frac{\partial}{\partial y}|f(x,y)|^p dy\, dx$$

$$= \int_X \int_{y=0}^{\infty} p|f(x,y)|^{p-1}\text{sign } f(x,y)\, \frac{\partial f}{\partial y}(x,y)\,dy\, dx$$

$$\leq C\left\{\int_X \int_{y=0}^{\infty} |f(x,y)|^p dy\, dx\right\}^{\frac{p-1}{p}} \left\{\int_X \int_{y=0}^{\infty} |D_y f(x,y)|^p dy\, dx\right\}^{\frac{1}{p}}$$

by Hölder's inequality. Thus

$$\| Bf\|_{L^p(X)} \leq C\| f\|^{(p-1)/p}_{L^p(X\times Y^+)} \| D_y f\|^{1/p}_{L^p(X\times Y^+)}$$

$$\| Bf\|_{L^p(X)} \leq C\| f\|^{(p-1)/p}_{L^p(X\times Y^+)} \| f\|^{1/p}_{L^p_{\sigma/\rho}(X\times Y^+)}.$$

Consider the partial smoothing operator

$$T_v f(x,y) = \frac{1}{v}\int_0^v f(x,y+w)\,dw.$$

Then $T_v : \mathcal{S}(X \times Y^+) \to \mathcal{S}(X \times Y^+)$ for $v > 0$. We have immediately that

$$\| T_v f\|_{L^p_k(X\times Y^+)} \leq C\| f\|_{L^p_k(X\times Y^+)}$$

for all k. Also

$$T_v f(x,y) = \frac{1}{v}\int_y^{y+v} f(x,z)\,dz$$

$$D_y T_v f(x,y) = \frac{1}{v}[f(x,y+v) - f(x,y)]$$

$$\| D_y T_v f\|_{L^p(X\times Y^+)} \leq C\, v^{-1}\| f\|_{L^p(X\times Y^+)}$$

$$\| D_y T_v f\|_{L^p(X\times Y^+)} \leq C\| f\|_{L^p_{\sigma/\rho}(X\times Y^+)}.$$

Then by interpolation for $0 \leq k \leq \sigma/\rho$

$$\| D_y T_v f \|_{L^p(X\times Y^+)} \leq C\, v^{\rho k/\sigma - 1} \| f \|_{L^p_k(X\times Y^+)} .$$

Also

$$[I - T_v] f(x,y) = \frac{1}{v} \int_{w=0}^{v} \int_{u=0}^{w} D_y f(x, y+u)\, du\, dw$$

$$\| [I - T_v] f \|_{L^p(X\times Y^+)} \leq C\, v \| D_y f \|_{L^p(X\times Y^+)}$$

$$\| [I - T_v] f \|_{L^p(X\times Y^+)} \leq C\, v \| f \|_{L^p_{\sigma/\rho}(X\times Y^+)}$$

$$\| [I - T_v] f \|_{L^p(X\times Y^+)} \leq C \| f \|_{L^p(X\times Y^+)} .$$

Then by interpolation for $0 \leq k \leq \sigma/\rho$

$$\| [I - T_v] f \|_{L^p(X\times Y^+)} \leq C\, v^{\rho k/\sigma} \| f \|_{L^p_k(X\times Y^+)} .$$

Thus for $0 \leq k \leq \sigma/\rho$

$$\| [T_{v/2} - T_v] f \|_{L^p(X\times Y^+)} \leq C\, v^{\rho k/\sigma} \| f \|_{L^p_k(X\times Y^+)}$$

$$\| D_y [T_{v/2} - T_v] f \|_{L^p(X\times Y^+)} \leq C\, v^{\rho k/\sigma - 1} \| f \|_{L^p_k(X\times Y^+)} .$$

Thus if $f \in \mathcal{S}(X \times Y^+)$

$$\| B [T_{v/2} - T_v] f \|_{L^p(X)} \leq C v^{\frac{\rho k}{\sigma} - \frac{1}{p}} \| f \|_{L^p_k(X\times Y^+)} .$$

Also

$$\| B\, T_1 f \|_{L^p(X)} \leq C \| f \|_{L^p_k(X\times Y^+)} .$$

So

$$\| B\, T_{2^{-K}} f \|_{L^p(X)} \leq \| B T_1 f \|_{L^p(X)} + \sum_{j=1}^{K} \| B [T_{2^{-j}} - T_{2^{-j+1}}] f \|_{L^p(X)}$$

and by the convergence of the geometric series, if $1/p < \frac{\rho k}{\sigma} \leq 1$

$$\| B T_{2^{-K}} f \|_{L^p(X)} \leq C \| f \|_{L^p_k(X \times Y^+)}$$

with a constant C independent of K. But $T_{2^{-K}} f \to f$ as $K \to \infty$. Thus

$$\| Bf \|_{L^p(X)} \leq C \| f \|_{L^p_k(X \times Y^+)} .$$

Thus B extends to a continuous linear map

$$B: L^p_k(X \times Y^+) \to L^p(X) \quad \text{if} \quad \frac{\rho k}{\sigma} > \frac{1}{p} .$$

We define the space

$$\partial L^p_k(X) = L^p_k(X \times Y^+)/\mathrm{Ker}\, B.$$

It is the space of "boundary values" of functions in $L^p_k(X \times Y^+)$. The norm is

$$\| g \|_{\partial L^p_k(X)} = \inf\{ \| f \|_{L^p_k(X \times Y^+)} : Bf = g \}.$$

The space $\partial L^p_k(X)$ is almost but not quite $L^p_{k-\sigma/\rho p}(X)$. The interested reader may consult Stein [22] for a reference. Note that since $\mathcal{S}(X \times Y^+/0)$ is dense, if $f \in L^p_k(X \times Y^+/0)$ then $Bf = 0$.

The operator $W(D_x)$ where

$$W(\xi) = (1 + \xi_1^{2\sigma_1} + \dots + \xi_n^{2\sigma_n})^{1/2\sigma}$$

commutes with the restriction to the boundary B. Therefore we have the following:

<u>Theorem</u>. If $0 \leq \ell < k - \sigma/\rho p$ then the restriction to the boundary defines a continuous linear map

$$B: L^p_k(X \times Y^+) \to L^p_\ell(X).$$

We also have the following:

Patching Theorem. Suppose $1 < p < \infty$ and $1/p < \rho k/\sigma < 1 + 1/p$. If $f_- \in L^p_k(X \times Y^-)$ and $f_+ \in L^p_k(X \times Y^+)$ agree on the boundary, so that $f_-|X = f_+|X$ in $L^p(X)$, then there exists a unique $f \in L^p_k(X \times Y)$ with $f|X \times Y^- = f_-$ and $f|X \times Y^+ = f_+$.

Proof. Let $f = f_-$ on $X \times Y^-$ and $f = f_+$ on $X \times Y^+$. This defines f as a distribution and in fact $f \in L^p(X \times Y)$. In order to show that $f \in L^p_k(X \times Y)$ we must show $W(D_x)^{\sigma/\rho}f$ and $D_y f$ belong to $L^p_{k-\sigma/\rho}(X \times Y)$. If $f_+ \in L^p_k(X \times Y^+)$ then $W(D_x)^{\sigma/\rho}f_+ \in L^p_{k-\sigma/\rho}(X \times Y^+)$. We know that

$$Z^*_+C^*_+ : L^p_{k-\sigma/\rho}(X \times Y^+/0) \to L^p_{k-\sigma/\rho}(X \times Y^+)$$

is an isomorphism since $-1 + 1/p < \rho(k-\sigma/\rho)/\sigma < 1/p$. The formula

$$\int_{X \times Y^+} f_+ \cdot W(D_x)^{\sigma/\rho} g = \int_{X \times Y^+} (Z^*_+C^*_+)^{-1} W(D_x)^{\sigma/\rho} f_+ \cdot g$$

is valid for all smooth f_+ and smooth g in $\mathcal{S}(X \times Y^+)$, and is well-defined for $f \in L^p_k(X \times Y^+)$ and $g \in \mathcal{S}(X \times Y^+)$; hence the formula remains valid in this larger context. Therefore for all $g \in \mathcal{S}(X \times Y)$

$$\int_{X \times Y} W(D_x)^{\sigma/\rho} f \cdot g = \int_{X \times Y} f \cdot W(D_x)^{\sigma/\rho} g$$

$$= \int_{X \times Y^+} f_+ \cdot W(D_x)^{\sigma/\rho} g + \int_{X \times Y^-} f_- \cdot W(D_x)^{\sigma/\rho} g$$

$$= \int_{X \times Y} \{(Z^*_+C^*_+)^{-1} W(D_x)^{\sigma/\rho} f_+ + (Z^*_-C^*_-)^{-1} W(D_x)^{\sigma/\rho} f_-\} g.$$

Therefore $W(D_x)^{\sigma/\rho} f \in L^p_{k-\sigma/\rho}(X \times Y)$.

Likewise the formula

$$\int_{X \times Y^+} f_+ \cdot D_y g = \int_{X \times Y^+} (Z^*_+C^*_+)^{-1} D_y f_+ \cdot g + \int_X f_+ \cdot g$$

is valid for all smooth f_+ and g in $\mathcal{S}(X \times Y^+)$, and is well-defined if $f_+ \in L^p_k(X \times Y^+)$.

Hence it remains valid in the larger context. Then we have

$$\int_{X\times Y} D_y f \cdot g = \int_{X\times Y} f \cdot D_y g = \int_{X\times Y^+} f_+ \cdot D_y g + \int_{X\times Y^-} f_- \cdot D_y g$$

$$= \int_{X\times Y} \{(Z_+^* C_+^*)^{-1} D_y f + (Z_-^* C_-^*)^{-1} D_y f_-\} g$$

since the boundary integrals $\int_X f_+ \cdot g = \int_X f_- \cdot g$ cancel. Therefore $D_y f \in L^p_{k-\sigma/\rho}(X \times Y)$ also. This completes the proof.

13. Let X be the space of $\{x_1, \dots, x_n\}$ with weights $\sigma_1, \dots, \sigma_n$ having least common multiple σ. Let $\mathcal{C}_0(X)$ denote the Banach space of continuous functions on X vanishing at ∞ with the supremum norm.

Sobolev Embedding Theorem. If $k > (\sigma/\sigma_1 + \dots + \sigma/\sigma_n)/p$ then there is a continuous inclusion $L^p_k(X) \subseteq \mathcal{C}_0(X)$.

Proof. The results of the previous section show that in restricting to the subspace $x_n = 0$ we lose just a little more than $\sigma/\sigma_n p$ derivatives. Thus by restricting one dimension at a time we obtain an estimate

$$|f(0)| \leq c\|f\|_{L^p_k(X)}$$

provided $k > (\sigma/\sigma_1 + \dots + \sigma/\sigma_n)1/p$. To be precise this holds for $f \in \mathcal{S}(X)$ which is dense in $L^p_k(X)$. For a general $f \in L^p_k(X)$ choose a sequence $f_m \in \mathcal{S}(X)$ converging to f. Then

$$\sup_x |f_m(x) - f_{m'}(x)| \leq c\| f_m - f_{m'} \|_{L^p_k(x)}$$

and f_m is Cauchy in $L^p_k(X)$ so f_m is also Cauchy in $\mathcal{C}_0(X)$. Therefore $f \in L^p_k(X)$ is equal (as a distribution in $\mathcal{S}(X)^*$) to a function in $\mathcal{C}_0(X)$ and the inclusion

$$L^p_k(X) \to \mathcal{C}_0(X)$$

is continuous.

14. By the same methods we can handle corners. Let $X \times Y \times Z$ denote the space of the variables $\{x_n, \ldots, x_n, y, z\}$ with the last two distinguished, and consider the four corners with y and z positive or negative. The operators Z, C, E and R operate separately in the variables y and z in each corner. Thus we obtain a commutative diagram of split exact sequences

$$\begin{array}{ccccccccc}
 & & 0 & & 0 & & 0 & & \\
 & & \downarrow\uparrow & & \downarrow\uparrow & & \downarrow\uparrow & & \\
0 & \rightleftarrows & \mathcal{S}(X\times Y^-/0\times Z^-/0) & \rightleftarrows & \mathcal{S}(X\times Y\times Z^-/0) & \rightleftarrows & \mathcal{S}(X\times Y^+\times Z^-/0) & \rightleftarrows & 0 \\
 & & \downarrow\uparrow & & Z^z_-\downarrow\uparrow R^z_+ & & \downarrow\uparrow & & \\
0 & \rightleftarrows & \mathcal{S}(X\times Y^-/0\times Z) & \underset{R^y_-}{\overset{Z^y_-}{\rightleftarrows}} & \mathcal{S}(X\times Y\times Z) & \underset{E^y_+}{\overset{C^y_+}{\rightleftarrows}} & \mathcal{S}(X\times Y^+\times Z) & \rightleftarrows & 0 \\
 & & \downarrow\uparrow & & C^z_+\downarrow\uparrow E^z_+ & & \downarrow\uparrow & & \\
0 & \rightleftarrows & \mathcal{S}(X\times Y^-/0\times Z^+) & \rightleftarrows & \mathcal{S}(X\times Y\times Z^+) & \rightleftarrows & \mathcal{S}(X\times Y^+\times Z^+) & \rightleftarrows & 0 \\
 & & \downarrow\uparrow & & \downarrow\uparrow & & \downarrow\uparrow & & \\
 & & 0 & & 0 & & 0 & &
\end{array}$$

By replacing $\mathcal{S}(X\times Y\times Z)$ with $L^p_n(X\times Y\times Z)$ we obtain the definition of the spaces $L^p_n(X\times Y^-/0\times Z^-/0)$, $L^p_n(X\times Y^+\times Z^-/0)$, $L^p_n(X\times Y^-/0\times Z^+)$ and $L^p_n(X\times Y^+\times Z^+)$. Similar remarks apply to corners in even more variables. In the most general situation we divide the variables into three groups x_i, y_j and z_k. Consider the corner with the x_i unrestricted, the $y_j \leq 0$ and the $z_k \geq 0$. The space $L^p_n(X\times Y^-/0\times Z^+)$ is defined to be those distributions in $\mathcal{S}(X\times Y^-\times Z^+/0)^*$ which are induced by an element of $L^p_n(X\times Y\times Z) \subseteq \mathcal{S}(X\times Y\times Z)^*$ whose restriction to $\mathcal{S}(X\times Y^+/0\times Z)$ is zero.

The following notation provides a transition to a coordinate-free description. Let $\mathcal{C}$ represent a corner $\{x_i, y_j \leq 0, z_k \geq 0\}$. The boundary $\partial\mathcal{C}$ is again a union of corners of lower dimension, which we call the faces of $\mathcal{C}$; they are the sets $\{y_1 = 0\}, \ldots, \{z_1 = 0\}, \ldots$. Let $\mathcal{Q}$ be the set of faces $\{y_1 = 0\}, \ldots, \{y_m = 0\}$. Then we write $L_n^p(\mathcal{C}/\mathcal{Q})$ for $L_n^p(X \times Y^-/0 \times Z^+)$. For any choice of faces $\mathcal{Q}$ such a representation is possible. If $\partial\mathcal{C} - \mathcal{Q}$ is the collection of faces not in $\mathcal{Q}$ then $L_{-n}^q(\mathcal{C}/\partial\mathcal{C} - \mathcal{Q})$ is the space dual to $L_n^p(\mathcal{C}/\mathcal{Q})$.

The extension and restriction operators provide a representation of $\mathcal{S}(\mathcal{C}/\mathcal{Q})$ as a direct summand of $\mathcal{S}(X \times Y \times Z)$ in a way that extends to a representation of $L_n^p(\mathcal{C}/\mathcal{Q})$ as a direct summand of $L_n^p(X \times Y \times Z)$ for every p and n. We then conclude that interpolating between $L_n^p(\mathcal{C}/\mathcal{Q})$ and $L_m^q(\mathcal{C}/\mathcal{Q})$ produces $L_k^r(\mathcal{C}/\mathcal{Q})$ from the following lemma:

Lemma. If A_0, A_1 and B_0, B_1 are two interpolation pairs then so is $A_0 \oplus B_0, A_1 \oplus B_1$, and interpolating between $A_0 \oplus B_0$ and $A_1 \oplus B_1$ produces $A_\theta \oplus B_\theta$.

Proof. Just note that

$$\mathcal{H}(A_0 \oplus B_0, A_1 \oplus B_1) = \mathcal{H}(A_0, A_1) \oplus \mathcal{H}(B_0, B_1).$$

15. Let X be a finite dimensional vector space. A foliation on X is an increasing family of subspaces $\{0\} \subseteq X_1 \subseteq \cdots \subseteq X_k \subseteq X$. If Y is another finite dimensional vector space with a foliation $\{0\} \subseteq Y_1 \subseteq \cdots \subseteq Y_k \subset Y$ and if $f: U \subseteq X \to V \subseteq Y$ is a smooth map we say f preserves the foliation if f maps each subspace $x + X_i$ into the subspace $f(x) + Y_i$. A foliated manifold is a manifold modeled on foliated vector spaces whose coordinate transition functions all preserve the foliation. If $\mathcal{C}$ is a corner in X

we say the corner is compatible with the foliation if we can choose a basis so that $X = \{x_1,\dots,x_n\}$ and each $X_\ell = \{x_1,\dots,x_{j_\ell}\}$ while the corner $\mathcal{C} = \{x : x_i \geq 0$ for $\forall i \in I\}$ where I is some index set $\subseteq \{1,\dots,n\}$. In this case we say $\mathcal{C}$ is a foliated corner. A foliated manifold with corners is a manifold whose coordinate charts are defined on foliated corners and whose coordinate transition functions preserve the foliation. Let $\mathcal{M}$ be a foliated manifold with corners. We say that a subset $\mathcal{A}$ of $\partial\mathcal{M}$ is a proper boundary part if in each coordinate chart $\mathcal{A}$ is a union of faces of the corner $\mathcal{C}$. There will then be a complementary proper boundary part which we write $\partial\mathcal{M} - \mathcal{A}$. (Note that it is larger than the set-theoretic difference.) We assign weights $\sigma_1 > \sigma_2 > \dots > \sigma_k$ to the foliation, and in local coordinates if $X = \{x_1,\dots x_n\}$ and $X_\ell = \{x_1,\dots,x_{j_\ell}\}$ we let $\{x_{j_{\ell-1}} + 1,\dots,x_{j_\ell}\}$ have weight σ_ℓ. We define σ to be the least common multiple of $\sigma_1,\dots,\sigma_k$. Let $\mathcal{C}$ be a corner in X and $\mathcal{D}$ a corner in Y, and let $q : U \subseteq \mathcal{C} \to V \subseteq \mathcal{D}$ be a smooth diffeomorphism of an open set in one corner into an open set in another. Let $\mathcal{A}$ be a collection of faces in $\mathcal{C}$ and $\mathcal{B}$ the corresponding faces in $\mathcal{D}$ under the action of φ. Let γ be a smooth function with compact support in U. Define a map $T : \mathcal{S}(\mathcal{D}/\mathcal{B}) \to \mathcal{S}(\mathcal{C}/\mathcal{A})$ by

$$Tf(x) = \gamma(x)\ f(\varphi x).$$

<u>Lemma</u>. For $1 < p < \infty$ and $-\infty < n < \infty$ the map T extends by completion to a continuous linear map $T : L^p_n(\mathcal{D}/\mathcal{B}) \to L^p_n(\mathcal{C}/\mathcal{A})$.

<u>Proof</u>. When n is an integer multiple of σ this can be verified directly. It then follows for the intermediate values of n by interpolation. To be specific, if n is a positive integer multiple of σ and $f \in \mathcal{S}(\mathcal{C}/\partial\mathcal{C} - \mathcal{A})^*$ then $f \in L^p_n(\mathcal{C}/\mathcal{A}) \iff D^\gamma f \in L^p(\mathcal{C})$ for $\|\gamma\| \leq n$; and if n is a negative integer multiple of σ and $f \in \mathcal{S}(\mathcal{C}/\partial\mathcal{C} - \mathcal{A})^*$ then

$f \in L_n^p(\mathcal{C}/\mathcal{A}) \Longleftrightarrow f = \Sigma D^\gamma g_\gamma$ with $g_\gamma \in L^p(\mathcal{C})$ for $\|\gamma\| \leq n$. Also if $i \leq j_\ell < k$ then $\partial y_k/\partial x_i = 0$, since the foliation is preserved. The rest of the verification is straightforward.

Let $\mathcal{M}$ be a compact foliated manifold with corners. Let $\varphi: U \subseteq \mathcal{M} \to V \subseteq \mathcal{C}$ be a coordinate chart and γ a smooth function with compact support in U. Let $\mathcal{A}$ be a proper part of the boundary of $\mathcal{M}$ and let $\mathcal{B}$ be the corresponding set of faces of $\mathcal{C}$ under φ. Define

$$T: \mathcal{S}(\mathcal{C}/\mathcal{B}) \to \mathcal{S}(\mathcal{M}/\mathcal{A})$$

$$\tilde{T}: \mathcal{S}(\mathcal{M}/\mathcal{A}) \to \mathcal{S}(\mathcal{C}/\mathcal{B})$$

by

$$Tf(x) = \gamma(x)\, f(\varphi x)$$

$$\tilde{T}g(z) = \gamma(\varphi^{-1}z)\, g(\varphi^{-1}z).$$

Theorem. For each $1 < p < \infty$ and each n there exists a unique Banach space $L_n^p(\mathcal{M}/\mathcal{A})$ such that each such T and $\tilde{T}$ extends by completion to continuous linear maps

$$T: L_n^p(\mathcal{C}/\mathcal{B}) \to L_n^p(\mathcal{M}/\mathcal{A})$$

and

$$\tilde{T}: L_n^p(\mathcal{M}/\mathcal{A}) \to L_n^p(\mathcal{C}/\mathcal{B}).$$

Proof. Let φ_i be a finite collection of coordinate charts whose domains cover $\mathcal{M}$ and let γ_i be a partition of unity subordinate to this cover. Define

$$\tilde{T}_i f(x) = \gamma_i(\varphi^{-1}x)\, f(\varphi^{-1}x).$$

Let

$$L_n^p(\mathcal{M}/\mathcal{A}) = \{f \in \mathcal{S}(\mathcal{M}/\partial\mathcal{M}-\mathcal{A})^* : \tilde{T}_i f \in L_n^p(\mathcal{C}_i/\mathcal{B}_i)\}.$$

Then using the preceding Lemma we can easily see that this definition is independent of the choice of charts and has all the required properties. Moreover let β_i be smooth functions with compact support in the domains of the φ_i which are equal to 1 on the support of the γ_i. Let $T_i f(x) = \beta_i(x)\, f(\varphi x)$. Then $\Sigma_i\, T_i \tilde{T}_i = I$. This provides a representation of $L^p_n(\mathcal{M}/\mathcal{A})$ as a direct summand of the direct sum $\oplus_i L^p_n(\mathcal{C}_i/\mathcal{B}_i)$. We conclude as before that interpolating between $L^p_n(\mathcal{M}/\mathcal{A})$ and $L^q_m(\mathcal{M}/\mathcal{A})$ produces $L^r_k(\mathcal{M}/\mathcal{A})$.

Using a partition of unity we have the Sobolev embedding Theorem for M.

Theorem. If $k > (\sigma/\sigma_1 + \ldots + \sigma/\sigma_n)/p$ then $L^p_k(M) \subseteq \mathcal{C}(M)$, the space of continuous functions on M, with a continuous inclusion.

16. Theorem. Let $\mathcal{M}$ be a compact foliated manifold with corners and $\mathcal{A}$ a proper part of the boundary. If $-\infty < k < n < \infty$ then the natural inclusion

$$L^p_n(\mathcal{M}/\mathcal{A}) \rightarrow L^p_k(\mathcal{M}/\mathcal{A})$$

is compact.

Proof. First suppose n is very large and $k = 0$. Then if $f \in L^p_n(\mathcal{M}/\mathcal{A})$ the Sobolev embedding theorem tells us that f and its first derivatives are continuous. By Ascoli's theorem the inclusion into the continuous functions is compact. Hence $L^p_n(\mathcal{M}/\mathcal{A}) \rightarrow L^p_0(\mathcal{M}/\mathcal{A})$ is compact.

Next we claim the result is also true for large n any $k < n$. This follows from the following lemma:

Lemma. If the inclusion $A_1 \rightarrow A_0$ is compact and if A_θ is the space obtained by holomorphic interpolation then the inclusion $A_1 \rightarrow A_\theta$ is compact for $\theta < 1$.

Proof. Let x_n be a bounded sequence in A_1. Then there exists a subsequence, which we will also call x_n, which converges in A_0. Then x_n is Cauchy in A_0. Moreover

$$\|x_n-x_m\|_{A_\theta} \le C\|x_n-x_m\|_{A_1}^{\theta}\ \|x_n-x_m\|_{A_0}^{1-\theta}.$$

Thus x_n is Cauchy in A_θ as well and hence converges in A_θ.

Finally, we observe that if r is an integral multiple of σ and if P_i is a sufficiently ample collection of partial differential operators of weight r (in local coordinate charts) then the map

$$\bigoplus_i L^p_{n+r}(\mathcal{M}/\mathcal{A}) \to L^p_n(\mathcal{M}/\mathcal{A})$$

given by $\{f_i\} \to \Sigma\, P_i f_i$ is surjective. Let n and k be arbitrary and choose r large enough. In the diagram

$$\begin{array}{ccc} \bigoplus_i L^p_{n+r}(\mathcal{M}/\mathcal{A}) & \to & L^p_n(\mathcal{M}/\mathcal{A}) \\ \downarrow & & \downarrow \\ \bigoplus_i L^p_{k+r}(\mathcal{M}/\mathcal{A}) & \to & L^p_k(\mathcal{M}/\mathcal{A}) \end{array}$$

the two horizontal maps are surjective and the left vertical map is compact so the right vertical map is compact also.

Part III: Semi-Elliptic and Parabolic Equations

1. Let X denote the space of the variables $\{x_1,\dots,x_n\}$. Let each variable x_i have integer weight σ_i and put σ equal to the least common multiple of the σ_i. An index $\alpha = (\alpha_1,\dots,\alpha_n)$ has weight

$$\|\alpha\| = (\sigma/\sigma_1)\alpha_1 + \dots + (\sigma/\sigma_n)\alpha_n.$$

A heterogeneous constant coefficient partial differential operator A of weight r has the form

$$A(D) = \sum_{\|\alpha\|=r} a_\alpha D^\alpha.$$

Its symbol is the polynomial

$$A(\xi) = \sum_{\|\alpha\|=r} a_\alpha \xi^\alpha .$$

The operator $A(\xi)$ is semi-elliptic if $A(\xi) \neq 0$ for all real $\xi \neq 0$. In this case r must be an integral multiple of σ. If

$$\|\xi\| = (\xi_1^{2\sigma_1} + \dots + \xi_n^{2\sigma_n})^{1/2\sigma}$$

then for some $\varepsilon > 0$

$$|A(\xi)| \geq \varepsilon\|\xi\|^r.$$

If $W(\xi)$ is the weight function

$$W(\xi) = (1+\xi_1^{2\sigma_1} + \dots + \xi_n^{2\sigma_n})^{1/2\sigma}$$

then $W(\xi)$ and $\|\xi\|$ are comparable for $\|\xi\| \geq 1$.

Let $\psi(\xi)$ be a smooth function with $\psi(\xi) = 1$ for $\|\xi\| \leq 1$ and $\psi(\xi) = 0$ for $\|\xi\| \leq 2$. Put $G(\xi) = [1-\psi(\xi)]/A(\xi)$ for $\|\xi\| \geq 1$ and $G(\xi) = 0$ for $\|\xi\| \leq 1$. Then $G(\xi)$ is smooth. Moreover it follows from Stein's multiplier theorem that $G(D)$ defines a continuous linear map

$$G(D):\ L^p_{n-r}(X) \to L^p_n(X)$$

for every real n. Moreover G is an approximate inverse for A in that

$$I - A(D)G(D) = \psi(D) = I - G(D)A(D)$$

and $\psi(D)$ defines a bounded linear map

$$\psi(D): L^p_{-m}(X) \to L^p_n(X)$$

for any real m and n.

2. In order to study boundary value problems we distinguish the last variable. Let $X \times Y$ denote the space of the variables $\{x_1,\ldots,x_n,y\}$ and write $Y^+ = \{y \geq 0\}$. We consider an elliptic homogeneous operator

$$A(D) = \sum_{\|(\alpha,\beta)\| = r} a_{\alpha\beta} D_x^\alpha D_y^\beta$$

where $x_1,\ldots,x_n$ have weights $\sigma_1,\ldots,\sigma_n$ and y has weight ρ, with σ the least common multiple of $\sigma_1,\ldots,\sigma_n,\rho$, and

$$\|(\alpha,\beta)\| = (\sigma/\sigma_1)\alpha_1 + \ldots + (\sigma/\sigma_n)\alpha_n + (\sigma/\rho)\beta.$$

The symbol of A is a polynomial

$$A(\xi,\eta) = \sum_{\|(\alpha,\beta)\| = r} a_{\alpha\beta}\xi^\alpha \eta^\beta$$

where $\xi_1,\ldots,\xi_n,\eta$ are the variables dual to $x_1,\ldots,x_n,y$. For every real $\xi \neq 0$ the polynomial $A(\xi,\eta)$ has no real zeros. If $\dim X \geq 2$ then $X-\{0\}$ is connected so the number m of zeros with positive imaginary part is constant; otherwise we must assume this. The coefficient of the highest order normal derivative $D_y^{r/\rho}$ is non-zero, so it is no loss to assume it is one. We then let $A^+(\xi,\eta)$ be the factor of $A(\xi,\eta)$ corresponding to the roots with $\operatorname{Im} \eta > 0$. If Γ is a path in the upper half plane enclosing these roots then we have an explicit formula

$$A^{+}(\xi,\eta) = \exp \int_{\in\Gamma} \log(\eta-\omega) \frac{\frac{\partial A}{\partial \omega}(\eta,\omega)}{A(\eta,\omega)} d\omega .$$

Therefore we see that

$$A^{+}(\xi,\eta) = \sum_{\beta=0}^{m} a^{+}_{\beta}(\xi)\eta^{\beta} .$$

where the $c^{+}_{\beta}(\xi)$ are heterogeneous functions of ξ of weight $(m-\beta)\rho$, where we say the function $h(\xi)$ is heterogeneous of weight w if h is defined and smooth for $\xi \neq 0$ and if for all $t > 0$

$$h(t^{\sigma/\sigma_1}\xi_1,\ldots,t^{\sigma/\sigma_n}\xi_n) = t^{w}h(\xi_1,\ldots,\xi_n).$$

We consider m heterogeneous boundary operators B^{j} $(1 \leq j \leq m)$ of weights r_j with

$$B^{j}(D) = \sum_{\|(\alpha,\beta)\|=r_j} b^{j}_{\alpha\beta} D^{\alpha}_{x} D^{\beta}_{y}$$

Their symbols are the polynomials

$$B^{j}(\xi,\eta) = \sum_{\|(\alpha,\beta)\|=r_j} b^{j}_{\alpha\beta} \xi^{\alpha} \eta^{\beta}.$$

We consider these as polynomials in η whose coefficients are heterogeneous functions (actually polynomials) of ξ.

$$B^{j}(\xi,\eta) = \Sigma b^{j}_{\beta}(\xi)\eta^{\beta}$$

where $b^{j}_{\beta}(\xi)$ is heterogeneous of weight $r_j-\rho\beta$. We reduce the B^{j} modulo A^{+} (as polynomials in η) and write

$$B^{j}(\xi,\eta) \equiv C^{j}(\xi,\eta) \bmod E^{+}(\xi,\eta)$$

with

$$C^{j}(\xi,\eta) = \sum_{\beta=0}^{m-1} c^{j}_{\beta}(\xi)\eta^{\beta}.$$

We assume that the boundary operators satisfy the complementary boundary condition

(CBC) The B^j are linearly independent modulo A^+ for all real $\xi \neq 0$ which is equivalent to the condition

$$\det c_\beta^j(\xi) \neq 0 \quad \text{for all real} \quad \xi \neq 0.$$

In this case we can find an inverse matrix of heterogeneous functions $e_k^\beta(\xi)$ with

$$c_\beta^j(\xi)\; e_k^\beta(\xi) = \delta_k^j.$$

3. Recall that

$$A^+(\xi,\eta) = \sum_{\beta=0}^{m} a_\beta^+ (\xi)\eta^\beta.$$

Let

$$A_\alpha^+(\xi,\eta) = \sum_{\beta=0}^{m-\alpha-1} a_{\alpha+\beta+1}^+(\xi)\eta^\beta$$

for $0 \leq \alpha \leq m-1$. Then if Γ is a path in the upper half plane enclosing the roots of A^+

$$\int_{\eta\in\Gamma} \frac{A_\alpha^+(\xi,\eta)}{A^+(\xi,\eta)}\, \eta^\beta \, d\eta = \delta_\alpha^\beta$$

for $0 \leq \beta \leq m-1$ by Cauchy residues, deforming the path Γ to ∞. Define for real $\xi \neq 0$ and $y \geq 0$

$$K_j(\xi,y) = \int_{\eta\in\Gamma} \sum_{\alpha=0}^{m-1} e_j^\alpha(\xi)\, \frac{A_\alpha^+(\xi,\eta)}{A^+(\xi,\eta)}\, e^{i\eta y}\, d\eta\ .$$

The kernel $K_j(\xi,y)$ is skew-heterogeneous of weight $-r_j$ in the sense that for $t > 0$

$$K_j(t^{\sigma/\sigma_1}\xi_1, \ldots, t^{\sigma/\sigma_n}\xi_n, t^{-\sigma/\rho}y)$$

$$= t^{-r_j} K_j(\xi_1,\ldots,\xi_n,y).$$

Moreover the same path Γ can be made to work for all real ξ with $\|\xi\| = 1$. If Γ has distance $\epsilon > 0$ from the real axis then

$$|K_j(\xi,y)| \leq Ce^{-\epsilon y} \quad \text{for} \quad \|\xi\| = 1.$$

Likewise $D_\xi^\alpha D_y^\beta K_j(\xi,y)$ is skew-heterogeneous of weight $-r_j - \|\alpha\| + (\sigma/\rho)\beta$ (with $\|\alpha\| = (\sigma/\sigma_1)\alpha_1+\ldots+(\sigma/\sigma_n)\alpha_n$) and satisfies $|D_\xi^\alpha D_y^\beta K_j(\xi,y)| \leq Ce^{-\epsilon y}$ for $\|\xi\| = 1$. Then for all $\xi \neq 0$

$$|\xi^\gamma y^\delta D_\xi^\alpha D_y^\beta K_j(\xi,y)| \leq C\|\xi\|^{-w} e^{-\epsilon\|\xi\|^{\sigma/\rho}y}$$

where $w = r_j + \|\alpha\| - \|\gamma\| + (\sigma/\rho)(\delta-\beta)$. Let $\psi(\xi)$ be a smooth function as before equal to 1 for $\|\xi\| \leq 1$ and 0 for $\|\xi\| \leq 2$. Define

$$H_j(\xi,y) = [1-\psi(\xi)]K_j(\xi,y),$$

for $\|\xi\| \geq 1$ and $H_j(\xi,y) = 0$ for $\|\xi\| \leq 1$. Then $H_j(\xi,y)$ is smooth and satisfies

$$|\xi^\gamma y^\delta D_\xi^\alpha D_y^\beta H_j(\xi,y)| \leq C\, W(\xi)^{-w} e^{-\epsilon W(\xi)^{\sigma/\rho}y}$$

since $\|\xi\|$ and $W(\xi)$ are comparable for $\|\xi\| \geq 1$. If $f \in \mathcal{S}(X \times Y^+)$ we define the partial Fourier transform $\tilde{f} \in \mathcal{S}(\Xi \times Y^+)$ by

$$\tilde{f}(\xi,y) = \int e^{-i\langle\xi,x\rangle} \tilde{f}(x,y)dx.$$

This is an isomorphism of $\mathcal{S}(X \times Y^+)$ onto $\mathcal{S}(\Xi \times Y^+)$. Define the operator

$$H_j(D_x,y): \mathcal{S}(X) \to \mathcal{S}(X \times Y^+)$$

by

$$\widetilde{H_j h}(\xi,y) = H_j(\xi,y)\tilde{h}(\xi).$$

4. Let $\operatorname{Ker} A \subseteq \mathcal{S}(X \times Y^+)$ be the kernel of $A: \mathcal{S}(X \times Y^+) \to \mathcal{S}(X \times Y^+)$. Let $\bigoplus_{j=1}^{m} \mathcal{S}(X)$ denote the direct sum of $\mathcal{S}(X)$ with itself m times. Then $Bf = (B^1 f,\ldots,B^m f)$ defines a map

$$B: \mathcal{S}(X \times Y^+) \to \bigoplus_{j=1}^{m} \mathcal{S}(X).$$

Likewise we define

$$H: \bigoplus_{j=1}^{m} \mathcal{S}(X) \to \mathcal{S}(X \times Y^+)$$

by

$$H(h_1,\ldots,h_m) = H_1 h_1 + \ldots + H_m h_m.$$

Then H is an approximate inverse to B on $\operatorname{Ker} A$ in the following sense:

<u>Theorem.</u> (1) $AH = 0$.

(2) $[I-BH]h = \psi(D_x)h$ for all $h \in \bigoplus_{j=1}^{m} \mathcal{S}(X)$.

(3) $[I-HB]f = \psi(D_x)f$ for all $f \in \operatorname{Ker} A$.

<u>Proof.</u> (1) Let $h = (h_1,\ldots,h_m)$. Then

$$\widetilde{Hh}(\xi,y) = \sum_{j=1}^{m} [1-\psi(\xi)]K_j(\xi,y)\tilde{h}_j(\xi)$$

$$\widetilde{AHh}(\xi,y) = A(\xi,D_y)\widetilde{Hh}(\xi,y) = \sum_{j=1}^{m} [1-\psi(\xi)]A(\xi,D_y)K_j(\xi,y)\tilde{h}_j(\xi).$$

Now $A(\xi,D_y)e^{i\eta y} = A(\xi,\eta)e^{i\eta y}$ so

$$\widetilde{AHh}(\xi,y) = \int_{\eta\in\Gamma} \sum_{j=1}^{m} \sum_{\alpha=0}^{m-1} [1-\psi(\xi)]e_j^{\alpha}(\xi)A(\xi,\eta)\frac{A_{\alpha}^{+}(\xi,\eta)}{A^{+}(\xi,\eta)}\tilde{h}_j(\xi)e^{i\eta y}d\eta \ .$$

But $A^{+}(\xi,\eta)$ is a factor of $A(\xi,\eta)$ so the integral is zero. Thus $AH = 0$.

(2) Likewise

$$\widetilde{B^k Hh}(\xi,y) = B^k(\xi,D_y)\widetilde{Hh}(\xi,y)$$

$$= \int_{\eta\in\Gamma} \sum_{j=1}^{m} \sum_{\alpha=0}^{m-1} [1-\psi(\xi)]e_j^{\alpha}(\xi)B^k(\xi,\eta)\frac{A_{\alpha}^{+}(\xi,\eta)}{A^{+}(\xi,\eta)}\tilde{h}_j(\xi)e^{i\eta y}d\eta \ .$$

But $B^k(\xi,\eta) \equiv C^k(\xi,\eta) = \sum_{\beta=0}^{m-1} c_{\beta}^{k}(\xi)\eta^{\beta}$ and $\int_{\eta\in\Gamma} \frac{A_{\alpha}^{+}(\xi,\eta)}{A^{+}(\xi,\eta)}\eta^{\beta}d\eta = \delta_{\alpha}^{\beta}$. Also $\sum_{\alpha=0}^{m-1} e_j^{\alpha}(\xi)c_{\alpha}^{k}(\xi) = \delta_j^k$. Thus $\widetilde{B^k Hh}(\xi,0) = [1-\psi(\xi)]\tilde{h}_k(\xi)$ so $BH = I-\psi(D_x)$ or $I - BH = \psi(D_x)$.

Finally (3) suppose $A \in \mathcal{S}(X\times Y^{+})$ and $Af = 0$. Let

$$g = HBf - f + \psi(D_x)f.$$

Then $Ag = AHBf - Af + A\psi(D_x)f = 0$. Also

$$Bg = BHBf - Bf + B\psi(D_x)f = [I-\psi(D_x)]Bf - [I-\psi(D_x)]Bf = 0$$

by part (2). Taking a partial Fourier transform

$$A(\xi,D_y)\tilde{g}(\xi,y) = 0 \quad \text{on } \Xi \times Y^{+}$$

$$B^k(\xi,D_y)\tilde{g}(\xi,0) = 0 \quad \text{on } \Xi \times \{0\}.$$

For $\xi \neq 0$ the complementing boundary condition assures that the only solution of this ordinary differential equation which is bounded on $X \times Y^{+}$ is zero. Thus $g = 0$. Therefore $[I-HB]f = \psi(D_x)f$ for $f \in \operatorname{Ker} A$.

5. We shall need a better approximate inverse G for A adapted to the boundary structure. Recall that in Section 8 of Part II we defined a smooth function $\varphi(x) \in (X)$ vanishing for $x \leq 0$ with

$$\int_0^\infty x^n \varphi(x)\,dx = (-1)^n \quad \text{all } \pm \text{ integers } n.$$

Let $\chi(y) = \varphi(y-1)$. Then $\chi(y) \in \mathcal{S}(Y)$ and $\chi(y) = 0$ for $y \leq 1$. Moreover

$$\int_1^\infty \chi(y)\,dy = 1$$

and for any positive integer n

$$\int_1^\infty y^n \chi(y)\,dy = \int_0^\infty (x+1)^n \varphi(x)\,dx = (1-1)^n = 0.$$

The Fourier transform $\hat{\chi}(\eta) \in \mathcal{S}(H)$ has therefore the same Taylor expansion at the origin as the constant function 1, so $1 - \hat{\chi}(\eta)$ has a zero of infinite order at the origin. Let $\psi(\xi)$ be as before a smooth function equal to 1 for $\|\xi\| \leq 1$ and 0 for $\|\xi\| \geq 2$. Put

$$\omega(\xi,\eta) = \psi(\xi)\hat{\chi}(\eta).$$

Then $\omega(D) = \omega(D_x, D_y) = \psi(D_x)\,\hat{\chi}(D_y)$. Now $\hat{\chi}(D_y)f = \chi * f$ and this convolution takes $\mathcal{S}(Y^-/0)$ into itself, and hence takes $\mathcal{S}(Y^+)$ into itself, since χ has support in $y \geq 1$. Thus $\omega(D): \mathcal{S}(X \times Y^+) \to \mathcal{S}(X \times T^+)$. In fact

$$\|\omega(D)f\|_{L^p_n(X\times Y^+)} \leq C\|f\|_{L^p_{-k}(X\times Y^+)}$$

for any n and k.

Moreover $1 - \omega(\xi,\eta)$ has a zero of infinite order at the origin so

$$G(\xi,\eta) = [1 - \omega(\xi,\eta)]/A(\xi,\eta)$$

is smooth even at the origin. Thus as before G defines a

continuous linear map of $\mathcal{S}(X \times Y) \to \mathcal{S}(X \times Y)$ or even

$$G(D): L^p_{n-r}(X \times Y) \to L^p_n(X \times Y)$$

and

$$I - AG = \omega(D) = I - GA.$$

Let $E: \mathcal{S}(X \times Y^+) \to \mathcal{S}(X \times Y)$ be an extension and $C: \mathcal{S}(X \times Y) \to \mathcal{S}(X \times Y^+)$ the cutoff as in Part II. Then $AC = CA$. Let

$$\tilde{G} = CGE: \mathcal{S}(X \times Y^+) \to \mathcal{S}(X \times Y^+).$$

Lemma. $I - A\tilde{G} = \omega(D)$.

Proof. $A\tilde{G} = ACGE = CAGE = C[I-\omega(D)]E = I-\omega(D)$ since $\omega(D)$ acts on $\mathcal{S}(X \times Y^+)$ by convolution independently of the extension.

In the other direction we have a weaker result.

Theorem. For all n and k

$$\|(I-\tilde{G}A)\psi(D_x)f\|_{L^p_n(X\times Y^+)} \leq C\|f\|_{L^p_{-k}(X\times Y^+)}.$$

Proof. It is enough to prove the theorem when n and k are integer multiples of σ/ρ. Since C and E commute with derivatives D_x in the X directions so does $\tilde{G}$. Thus

$$D_x^\alpha(I-\tilde{G}A)\psi(D_x)f = (I-\tilde{G}A)D_x^\alpha\psi(D_x)f.$$

But $\varphi(\xi)$ has compact support, so $\xi^\alpha\varphi(\xi)$ does also. Therefore these derivatives are easily estimated. On the other hand

$$A(I-\tilde{G}A)\psi(D_x)f = A\omega(D)\psi(D_x)f.$$

But the coefficient in A of the highest y derivative $D_y^{\rho r/\sigma}$ is not zero, so it is no loss to assume it is one. Write

$$A = D_y^{\rho r/\sigma} + A^\#$$

when $A^\#$ is also heterogeneous of weight r but involves only derivatives D_y^{β} for $\beta \leq (\rho r/\sigma)-1$

$$D_y^{\rho r/\sigma}(I - \tilde{G}A)\psi(D_x)f = -A^{\#}(I-\tilde{G}A)\psi(D_x)f + A\omega(D)\psi(D_x)f$$

This allows us to estimate y-derivatives in terms of x-derivatives, proceding by induction on k through integer multiples of σ/ρ.

6. Let $\mathcal{C}: \mathcal{S}(X\times Y^+) \to \mathcal{S}(X\times Y^+) \overset{m}{\underset{j=1}{\oplus}} \mathcal{S}(X)$ be defined by $\mathcal{C}f = (Af,Bf)$. Let $\mathcal{J}: \mathcal{S}(X\times Y^+) \overset{m}{\underset{j=1}{\oplus}} \mathcal{S}(X) \to \mathcal{S}(X\times Y^+)$ be defined by

$$\mathcal{J}(g,h) = \tilde{G}g + H(h-B\tilde{G}g).$$

Then $\mathcal{J}$ is an approximate inverse for $\mathcal{C}$ in the following sense. First we have $\mathcal{C}\mathcal{J}(g,h) = (A\tilde{G}g+AH(h-B\tilde{G}g),\ B\tilde{G}g+BH(h-B\tilde{G}g))$. But $A\tilde{G} = I-\omega(D)$ and $BH = I - \psi(D_x)$, while $AH = 0$. Therefore

$$\mathcal{C}\mathcal{J}(g,h) = (g-\omega(D)g,\ B\tilde{G}g + h - B\tilde{G}g - \psi(D_x)(h-B\tilde{G}g))$$

$$[I-\mathcal{C}\mathcal{J}](g,h) = (\omega(D)g,\ \psi(D_x)(h-B\tilde{G}g)).$$

In the other direction,

$$\mathcal{J}\mathcal{C}f = \tilde{G}Af + H(Bf-B\tilde{G}Af).$$

Now

$$A(f-\tilde{G}Af) = Af-A\tilde{G}Af = Af-[I-\omega(D)]Af = \omega(D)Af = A\omega(D)f.$$

Therefore $A(f-\tilde{G}Af-\omega(D)f) = 0$, so $f - \tilde{G}Af - \omega(D)f$ lies in the kernel of A. Then

$$HB(f-\tilde{G}Af-\omega(D)f) = [I-\psi(D_x)](f-\tilde{G}Af-\omega(D)f)$$

$$HB(f-\tilde{G}Af) = f - \tilde{G}Af - \omega(D)f - \psi(D_x)f + \psi(D_x)\tilde{G}Af + \psi(D_x)\omega(D)f + HB\omega(D)f.$$

$$[I-\mathcal{J}\mathcal{C}]f = \omega(D)f + \psi(D_x)[I-\tilde{G}A]f - \omega(D)\psi(D_x)f - HB\omega(D)f.$$

$$[I-\mathcal{J}\mathcal{C}]f = \psi(D_x)[I-\tilde{G}A]f + [I-HB-\psi(D_x)]\omega(D)f.$$

We summarize these results.

Theorem.

$$[I-\mathcal{C}\mathcal{J}](g,h) = (\omega(D)g, \psi(D_x)(h-B\tilde{G}g))$$

$$[I-\mathcal{J}\mathcal{C}]f = \psi(D_x)[I-\tilde{G}A]f + [I-HB-\psi(D_x)]\omega(D)f.$$

7. We know from Section 12 of Part II that restriction to the boundary defines a continuous linear map

$$B: L_n^p(X \times Y^+) \to \partial L_n^p(X)$$

when $n > \sigma/\rho p$. Therefore the boundary operator B^j of weight r_j defines a continuous linear map

$$B^j: L_n^p(X \times Y^+) \to \partial L_{n-r_j}^p(X)$$

when $n > r_j + \sigma/\rho p$. Now we show the H_j are coercive.

Theorem. If $n > r_j + \sigma/\rho p$ then H_j defines a continuous linear map

$$H_j: \partial L_{n-r_j}^p(X) \to L_n^p(X \times Y^+).$$

Proof. We shall prove this when $n - \sigma/\rho p$ is not an integer multiple of σ/ρ; the remaining cases then follow by interpolation. Let k be the first integer larger than $\rho n/\sigma - 1/p$. Then $n - \sigma/\rho p = (\sigma/\rho)(k-1+\alpha)$ with $0 < \alpha < 1$. In order to estimate $H_j h$ in $L_n^p(X \times Y^+)$ it is enough to estimate $W(D_x)^{\sigma k/\rho} H_j h$ and $D_y^k H_j h$ in $L_\gamma^p(X \times Y^+)$ where $\gamma = (\sigma/\rho)(-1+1/p+\alpha)$, so $-1+1/p < \rho\gamma/\sigma < 1/p$. Now $h \in \partial L_{n-r_j}^p(X)$ so we can write $h = W(D_x)^{r_j-(\sigma/\rho)(k-1)} \ell$ with $\ell \in \partial L_{\gamma+\sigma/\rho}^p(X)$. Then we must estimate $W(D_x)^{r_j+\sigma/\rho} H_j \ell$ and $D_y^k W(D_x)^{r_j-(\sigma/\rho)(k-1)} H_j \ell$ in $L_\gamma^p(X \times Y^+)$ in terms of $\ell \in \partial L_{\gamma+\sigma/\rho}^p(X)$. Let us write

$$H_j^{(1)}(\xi,y) = W(\xi)^{r_j+\sigma/\rho} H_j(\xi,y)$$

$$H_j^{(2)}(\xi,y) = W(\xi)^{r_j+\sigma/\rho-(\sigma/\rho)k} D_y^k H_j(\xi,y).$$

We must estimate $H_j^{(1)}\ell$ and $H_j^{(2)}\ell$ in $L_\gamma^p(X \times Y^+)$ for $\ell \in \partial L^p_{\gamma+\sigma/\rho}(X)$. We have estimates on $H_j(\xi,y)$ saying

$$|\xi^\gamma y^\delta D_\xi^\alpha D_y^\beta H_j(\xi,y)| \leq C\ W(\xi)^{-w}\ e^{-\epsilon W(\xi)^{\sigma/\rho} y}$$

with $w = r_j + \|\alpha\| - \|\gamma\| + (\sigma/\rho)(\delta-\beta)$. It follows that $H_j^{(1)}$ and $H_j^{(2)}$ satisfy

$$|\xi^\alpha y^\beta D_\xi^\alpha D_y^\beta H_j^{(i)}(\xi,y)| \leq C\ W(\xi)^{\sigma/\rho} e^{-\epsilon W(\xi)^{\sigma/\rho} y}\ .$$

If we integrate this with respect to y

$$\int_0^\infty |\xi^\alpha y^\beta D_\xi^\alpha D_y^\beta H_j^{(i)}(\xi,y)|\,dy \leq C$$

for $i = 1,2$. Therefore the $H_j^{(i)}$ satisfy the hypotheses of the following theorem which will complete this proof by showing

$$\| H_j^{(i)}\ell\|_{L_\gamma^p(X \times Y^+)} \leq C\|\ell\|_{\partial L^p_{\gamma+\sigma/\rho}(X)}.$$

Let $H(\xi,y)$ be a smooth function in ξ and y for $y \geq 0$ which is rapidly decreasing in y and slowly increasing in ξ. This means that for all i, j and α there is an m and a constant C with

$$|y^i D_\xi^\alpha D_y^j\ H(\xi,y)| \leq C(1 + |\xi|)^m.$$

Then we can define a map

$$H: \mathcal{S}(X) \to \mathcal{S}(X \times Y^+)$$

using the partial Fourier transform by the formula

$$\widetilde{Hh}(\xi,y) = H(\xi,y)\tilde{h}(\xi).$$

<u>Boundary Multiplier Theorem</u>. Suppose that for all indices α, β we have

$$\int_0^\infty |\xi^\alpha y^\beta D_\xi^\alpha D_y^\beta H(\xi,y)|\,dy \leq C$$

with $C = C(\alpha,\beta)$ independent of ξ. Then H extends to a bounded linear map

$$H: \partial L^p_{\gamma+\sigma/\rho}(X) \to L^p_\gamma(X \times Y^+)$$

for $(\sigma/\rho)(-1 + 1/p) < \gamma < \sigma/\rho p$.

Proof. We must show that for $h \in \mathcal{S}(X)$ we have an estimate

$$\| H\, h \|_{L^p_\gamma(X\times Y^+)} \leq C \| h \|_{\partial L^p_{\gamma+\sigma/\rho}(X)} .$$

If $h \in \mathcal{S}(X)$ we can write $h(x) = h(x,0)$ for some $h \in \mathcal{S}(X \times Y^+)$ with

$$\| h \|_{L^p_{\gamma+\sigma/\rho}(X\times Y^+)} \leq C \| h \|_{\partial L^p_{\gamma+\sigma/\rho}(X)} .$$

Then $\widetilde{H_j h}(\xi,y) = H_j(\xi,y)\tilde{h}(\xi,0)$. Following Agmon, Douglis and Nirenberg I [1] we write

$$\widetilde{H_j h}(\xi,y) = -\int_0^\infty \frac{\partial}{\partial s}\{H_j(\xi,y+s)\tilde{h}(\xi,s)\}ds$$

using the fundamental theorem of calculus.

The derivative of the product has two terms

$$H_j(\xi,y+s)\frac{\partial\tilde{h}}{\partial s}(\xi,s)$$

and

$$\frac{\partial H_j}{\partial s}(\xi,y+s)\tilde{h}(\xi,s).$$

For the first we have

$$\| \frac{\partial h}{\partial y} \|_{L^p_\gamma(X\times Y^+)} \leq C \| h \|_{\partial L^p_{\gamma+\sigma/\rho}(X)} .$$

The second term we write as

$$\{W(\xi)^{-\sigma/\rho} D_y H_j(\xi,y)\}\{W(\xi)^{\sigma/\rho}\tilde{h}(\xi)\}$$

and we also have

$$\| W(D_x)^{\sigma/\rho} h \|_{L^p_\gamma(X\times Y^+)} \leq C \| h \|_{\partial L^p_{\gamma+\sigma/\rho}(X)} .$$

Therefore both contributions are of the form

$$\int_0^\infty K(\xi, y+s) \ell(s) ds$$

where $K(\xi,y)$ is smooth, rapidly decreasing in y and slowly growing in ξ, and satisfies the estimates

$$\int_0^\infty |\xi^\alpha y^\beta D_\xi^\alpha D_y^\beta K(\xi,y)| dy \leq C$$

for all α and β, and where

$$\| \ell \|_{L^p_\gamma(X\times Y^+)} \leq C \| h \|_{\partial L^p_{\gamma+\sigma/\rho}(X)} .$$

Therefore the proof is completed by the following Lemma.

<u>Lemma</u>. Let $K(\xi,y)$ be smooth, rapidly decreasing in y and slowly growing in ξ. Define a map

$$K: \mathcal{S}(X \times Y^+) \to \mathcal{S}(X \times Y^+)$$

using the partial Fourier transform by

$$\widetilde{K\ell}(\xi,y) = \int_0^\infty K(\xi, y+s) \tilde{\ell}(\xi, s) ds .$$

If $K(\xi,y)$ satisfies the estimate

$$\int_0^\infty |\xi^\alpha y^\beta D_\xi^\alpha D_y^\beta K(\xi,y)| \leq C$$

for all α and β then K extends to a bounded linear map

$$K: L^p_\gamma(X \times Y^+) \to L^p_\gamma(X \times Y^+)$$

for $(\sigma/\rho)(-1+1/p) < \gamma < \sigma/\rho p$.

<u>Proof</u>. For γ in the indicated range we have $L^p_\gamma(X \times Y^+) = L^p_\gamma(X \times Y^+/0)$. Hence the zero extension $Z\ell$ is a

bounded extension of ℓ. Let EK be a smooth extension of K using the operator E constructed before. It is easy to verify that EK satisfies

$$\int_{-\infty}^{\infty} |\xi^\alpha y^\beta D_\xi^\alpha D_y^\beta EK(\xi,y)|\,dy \leq C.$$

Let $\widehat{EK}(\xi,\eta)$ denote the Fourier transform of EK in the y direction, and let $\widehat{Z\ell}(\xi,\eta)$ denote the total Fourier transform of $Z\ell$, which is the Fourier transform of $Z\widetilde{\ell}(\xi,y)$ in the y-direction. Then define a function $EK\ell$ by

$$\widetilde{EK\ell}(\xi,y) = \int_0^{\infty} EK(\xi,y+s)\widetilde{\ell}(\xi,s)\,ds$$

$$= \int_{-\infty}^{\infty} EK(\xi,y+s)Z\widetilde{\ell}(\xi,s)\,ds.$$

Then $\widetilde{EK\ell}(\xi,y) = \widetilde{K\ell}(\xi,y)$ for $y \geq 0$ so $EK\ell$ is an extension of $K\ell$. Moreover the Fourier transform takes convolution into multiplication so

$$\widehat{EK\ell}(\xi,\eta) = \widehat{EK}(\xi,\eta)\widehat{Z\ell}(\xi,\eta).$$

The estimate for EK implies

$$|\xi^\alpha \eta^\beta D_\xi^\alpha D_\eta^\beta \widehat{EK}(\xi,\eta)| \leq C$$

since the Fourier transform of an L^1 estimate is an L^∞ estimate. There are some commutators to watch since $\eta^\beta D_\eta^\beta$ transforms to $D_y^\beta y^\beta$, but $D_y \cdot y = y\,D_y + 1$ so we get a sum of terms $y^\gamma D_y^\gamma$ with $\gamma \leq \beta$. Therefore $\widehat{EK}(\xi,\eta)$ satisfies the hypotheses of Stein's multiplier theorem. It follows that

$$\|EK\ell\|_{L^p_\gamma(X\times Y)} \leq C\|Z\ell\|_{L^p_\gamma(X\times Y)}.$$

Therefore

$$\|K\ell\|_{L^p_\gamma(X\times Y^+)} \leq c\|EK\ell\|_{L^p_\gamma(X\times Y)}$$

$$\leq c\|Z\ell\|_{L^p_\gamma(X\times Y)} \leq c\|\ell\|_{L^p_\gamma(X\times Y^+/0)} \leq c\|\ell\|_{L^p_\gamma(X\times Y^+)}$$

for γ in the indicated range.

8. Let E,F,G,H be four Banach spaces and ℓ,m,p,q four continuous linear maps such that the diagram below commutes, which means $p\ell = qm$.

$$\text{(D)} \qquad \begin{array}{ccc} E & \xrightarrow{\ \ell\ } & F \\ {\scriptstyle m}\downarrow & & \downarrow{\scriptstyle p} \\ G & \xrightarrow[\ q\ ]{} & H \end{array} .$$

We say that (D) is an exact square if the following associated sequence is exact:

$$0 \to E \xrightarrow{\ \ell\oplus m\ } F \oplus G \xrightarrow{\ p\ominus q\ } H \to 0$$

where $\ell \oplus m(x) = (\ell x, mx)$ and $p \ominus q(y,z) = py-qz$.

We can compose exact squares.

<u>Theorem</u>. If

$$\begin{array}{ccc} E & \xrightarrow{\ \ell\ } & F \\ {\scriptstyle m}\downarrow & & \downarrow{\scriptstyle p} \\ G & \xrightarrow[\ q\ ]{} & H \end{array} \qquad \text{and} \qquad \begin{array}{ccc} F & \xrightarrow{\ r\ } & J \\ {\scriptstyle p}\downarrow & & \downarrow{\scriptstyle s} \\ H & \xrightarrow[\ t\ ]{} & K \end{array}$$

are exact squares so is the composition

$$\begin{array}{ccc} E & \xrightarrow{\ r\ell\ } & J \\ {\scriptstyle m}\downarrow & & \downarrow{\scriptstyle s} \\ G & \xrightarrow{\ tq\ } & K \end{array} .$$

Proof. This follows from an elementary diagram chase.

An easy method to prove that a square is exact is to construct a splitting. A splitting for the square (D) is a collection of four continuous linear maps ℓ', m', p', q' such that the diagram

$$\begin{array}{ccc} E & \underset{\ell}{\overset{\ell'}{\rightleftarrows}} & F \\ m' \uparrow\downarrow m & & p \downarrow\uparrow p' \\ G & \underset{q'}{\overset{q}{\rightleftarrows}} & H \end{array}$$

commutes in four ways

$$p\ell = qm \qquad\qquad m\ell' = q'p$$

$$\ell m' = p'q \qquad\qquad m'q' = \ell'p'$$

and splits in four ways

$$\ell'\ell + m'm = I \qquad\qquad \ell\ell' + p'p = I$$

$$mm' + q'q = I \qquad\qquad pp' + qq' = I.$$

A splitting for the square D defines a splitting for the associated sequence. If E, F, G, H are Hilbert spaces then every exact square splits.

<u>Theorem</u>. If A is a semi-elliptic operator of weight r with constant coefficients and $-\infty < k < n < \infty$ then the square

$$\begin{array}{ccc} L^p_n(X) & \xrightarrow{A} & L^p_{n-r}(X) \\ i \downarrow & & \downarrow i \\ L^p_k(X) & \xrightarrow{A} & L^p_{k-r}(X) \end{array}$$

is exact.

Note that this says three classical results at once:

(1) The map $L^p_n(X) \xrightarrow{A \oplus i} L^p_{n-r}(X) \oplus L^p_k(X)$ is one-to-one with closed

range. This happens if and only if for some C

$$\|f\|_{L^p_n(X)} \leq C\left(\|Af\|_{L^p_{n-r}(X)} + \|f\|_{L^p_k(X)}\right)$$

which is Gårding's inequality.

(2) If $f \in L^p_k(X)$ and $Af \in L^p_{n-r}(X)$ then $f \in L^p_n(X)$, which is the regularity theorem.

(3) If $g \in L^p_{k-r}(X)$ then there exists $f \in L^p_k(X)$ with $Af - g \in L^p_{n-r}(X)$, which is the existence of an approximate solution.

Proof. We construct a splitting for the diagram using the approximate inverse G of Section 1. In fact the results there show that a splitting is given by

$$\begin{array}{ccc}
L^p_n(X) & \underset{A}{\overset{G}{\rightleftarrows}} & L^p_{n-r}(X) \\
\psi(D)\uparrow\ \downarrow i & & i \downarrow\ \uparrow \psi(D) \\
L^p_k(X) & \underset{G}{\overset{A}{\rightleftarrows}} & L^p_{k-r}(X)
\end{array} .$$

Likewise for boundary value problems we have the following result:

Theorem. Let A be a semi-elliptic operator of weight r with constant coefficients and let B^j $(1 \leq j \leq m)$ be complementing boundary operators with constant coefficients with weights r_j. If $\sigma/\rho p + \max r_j < k < n < \infty$ then the following square is exact

$$\begin{array}{ccc}
L^p_n(X \times Y^+) & \xrightarrow{\ \mathcal{C}\ } & L^p_{n-r}(X \times Y^+) \ \bigoplus_{j=1}^{m} \partial L^p_{n=r_j}(X) \\
\Big\downarrow i & & \Big\downarrow i \\
L^p_k(X \times Y^+) & \xrightarrow{\ \mathcal{C}\ } & L^p_{k-r}(X \times Y^+) \ \bigoplus_{j=1}^{m} \partial L^p_{k-r_j}(X)
\end{array}$$

where $\mathcal{C}f = (Af, B^1f, \ldots, B^m f)$.

<u>Proof.</u> A splitting is given by the operator $\mathcal{G}$ defined in Section 6, as follows

$$\begin{array}{ccc}
L^p_n(X \times Y^+) & \underset{\mathcal{C}}{\overset{\mathcal{G}}{\rightleftarrows}} & L^p_{n-r}(X \times Y^+) \ \bigoplus_{j=1}^{m} \partial L^p_{n-r_j}(X) \\
\kappa \Big\uparrow \ \Big\downarrow i & & i \Big\downarrow \ \Big\uparrow \lambda \\
L^p_k(X \times Y^+) & \underset{\mathcal{G}}{\overset{\mathcal{C}}{\rightleftarrows}} & L^p_{k-r}(X \times Y^+) \ \bigoplus_{j=1}^{m} \partial L^p_{k-r_j}(X)
\end{array}$$

where

$$\kappa(f) = \psi(D_x)[I-\tilde{G}A]f + [I-HB-\psi(D_x)]\omega(D)f$$

$$\lambda(g,h) = (\omega(D)g,\ \psi(D_x)(h - B\tilde{G}g))$$

$$\mathcal{G}(g,h) = \tilde{G}g + H(h - B\tilde{G}g)$$

We know for any n and k that $\omega(D)$ maps $L^p_k(X \times Y^+)$ continuously into $L^p_n(X \times Y^+)$, and so does $\psi(D_x)\ [I - \tilde{G}A]$ by a result in Section 5. Likewise $\psi(D_x)$ maps $\partial L^p_k(X)$ continuously into $\partial L^p_n(X)$ for any n and k. Moreover $\mathcal{G}$ is continuous by the results of Section 7. We have already shown in Section 6 that

$$I - \mathcal{G}\mathcal{C} = \kappa \quad \text{and} \quad I - \mathcal{C}\mathcal{G} = \lambda.$$

A straightforward calculation shows that

$$i\mathcal{J} = \mathcal{J}i \qquad \mathcal{C}\kappa = \lambda\mathcal{C} \qquad \kappa\mathcal{J} = \mathcal{J}\lambda.$$

Therefore the square splits and so it is exact.

9. It is well known that for Banach spaces any linear map close enough to an invertible one is also invertible. We prove an analogous result for exact squares. Let E, F, G, H be Banach spaces. Then the set of commutative squares $SQ(E,F,G,H)$ is the closed subspace of

$$L(E,F) \times L(E,G) \times L(F,H) \times L(G,H)$$

consisting of those maps (ℓ,m,p,q) with $p\ell = qm$ in the diagram

$$\begin{array}{ccc} E & \xrightarrow{\ell} & F \\ {\scriptstyle m}\downarrow & & \downarrow{\scriptstyle p} \\ G & \xrightarrow[q]{} & H \end{array}$$

<u>Theorem.</u> The exact squares are relatively open in $SQ(E,F,G,H)$.

<u>Proof.</u> The square is exact if and only if the sequence

$$0 \to E \xrightarrow{\ell \oplus m} F \oplus G \xrightarrow{p \ominus q} H \to 0$$

is exact. Therefore the theorem reduces to a statement about exact sequences, so it follows from the next result.

If E, F, G are Banach spaces we let $S(E,F,G)$ denote the closed subspace of $L(E,F) \times L(F,G)$ consisting of those maps (ℓ,m) with $m\ell = 0$.

<u>Theorem.</u> The exact sequences

$$0 \to E \xrightarrow{\ell} F \xrightarrow{m} G \to 0$$

are relatively open in $S(E,F,G)$.

<u>Proof.</u> Suppose that

$$0 \to E \xrightarrow{\ell_0} F \xrightarrow{m_0} G \to 0$$

is exact.

(1) Since ℓ_0 is one-to-one with closed range, we can find a constant C so that $\|x\| \leq C\|\ell_0 x\|$. Suppose $\|\ell-\ell_0\| \leq 1/(2C)$. Then

$$\|x\| \leq C(\|\ell x\| + \|(\ell-\ell_0)x\|)$$

$$\|x\| \leq C\|\ell x\| + 1/2\|x\|$$

$$\|x\| \leq 2C\|\ell x\|$$

so ℓ is also one-to-one with closed range.

(2) Since m_0 is onto, we can find a constant C such that

$$\forall y \in G \quad \exists x \in F \quad \text{with } \|x\| \leq C\|y\| \text{ and } m_0 x = y.$$

Suppose $\|m-m_0\| \leq 1/(2C)$. Then $\|mx-y\| \leq 1/2\|y\|$. Thus by Lemma (b) in Part II, Section 5, we know m is onto also.

(3) The map $F/\text{Ker } m_0 \to G$ is an isomorphism. Therefore for some constant C, $\|y/\text{Ker } m_0\| \leq C\|m_0 y\|$. However, $\|y/\text{Ker } m_0\| = \inf\{\|y-z\| : z \in \text{Ker } m_0\}$. Therefore (with a slightly larger C)

$$\forall y \in F \quad \exists z \in \text{Ker } m_0 \quad \|y-z\| \leq C\|m_0 y\|.$$

However, $\text{Im}\ell_0 = \text{Ker } m_0$, and the map

$$E \to \text{Im } \ell_0 \subseteq F$$

is surjective. Therefore

$$\forall z \in \text{Ker } m_0 \quad \exists x \in E \quad \text{with } \ell_0 x = z \text{ and } \|x\| \leq C\|z\|.$$

But $\|z\| \leq \|z-y\| + \|y\| \leq C\|y\|$. This shows that for some constant C

$$\forall y \in F \;\; \exists x \in E \quad \text{with } \|x\| \leq C\|y\| \quad \|y-\ell_0 x\| \leq C\|m_0 y\|.$$

Suppose $\|\ell-\ell_0\| \leq 1/(4C)$ and $\|m-m_0\| \leq 1/(4C)$. Then
$\|y-\ell x\| \leq \|y-\ell_0 x\| + \|\ell_0-\ell\|\cdot\|x\|$ and
$\|y-\ell_0 x\| \leq C\|m_0 y\| \leq C\|my\| + C\|m-m_0\|\cdot\|y\|$. Thus
$\forall y \in F \;\; \exists x \in E$ with $\|x\| \leq C\|y\|$ and $\|y-\ell x\| \leq C\|my\| + 1/2\|y\|$.

Let $y_0 \in F$ with $my_0 = 0$. Choose inductively $x_0, y_1, x_1, y_2, x_2, \ldots$ where x_n is chosen as above with $\|x_n\| \leq C\|y_n\|$ and $\|y_n-\ell x_n\| \leq C\|my_n\| + 1/2\|y_n\|$, and where

$y_{n+1} = y_n - \ell x_n$. Then if $m\ell = 0$ we have

$my_n = my_{n-1} = \cdots = my_0 = 0 \qquad \|y_{n+1}\| = \|y_n - \ell x_n\| \leq 1/2\|y_n\|$.

Thus $y_n \to 0$ and $\Sigma\|y_n\| < \infty$. Then $x_n \to 0$ and $\Sigma\|x_n\| \leq C\Sigma\|y_n\| < \infty$. Let $x = \sum_{n=0}^{\infty} x_n$. Then

$$\ell x = \sum_{n=0}^{\infty} \ell x_n = \sum_{n=0}^{\infty} y_n - y_{n+1} = y_0.$$

Therefore $\text{Im}\ \ell = \text{Ker}\ m$ so the new sequence is exact also.

Now to obtain the corresponding result for exact squares we only need to observe that the map

$$SQ(E,F,G,H) \to S(E,F\oplus G,H),$$

associating to each square the corresponding sequence, is continuous, and the inverse image of an open set is open.

10. The stability of exact squares enables us to extend the preceding results to variable coefficients. It is convenient to work on small boxes with sides parallel to the coordinate axes. Such a box is a particularly simple example of a foliated manifold with corners. We adopt the following notation. Let $\mathcal{B}_\epsilon$ be the box

$$\mathcal{B}_\epsilon = \{(x_1,\ldots,x_n): |x_1| \leq \epsilon^{\sigma/\sigma_1},\ldots,|x_n| \leq \epsilon^{\sigma/\sigma_n}\}.$$

For boundary value problems we use the box

$$\mathcal{B}_\epsilon^+ = \{(x_1,\ldots,x_n,y): |x_1| \leq \epsilon^{\sigma/\sigma_1},\ldots,|x_n| \leq \epsilon^{\sigma/\sigma_n},$$
$$0 \leq y \leq \epsilon^{\sigma/\rho}\}.$$

We distinguish two parts of the boundary; $\partial_0\mathcal{B}_\epsilon^+$ is the part where $y = 0$ and $\partial_e\mathcal{B}_\epsilon^+$ is the exterior boundary complementary to $\partial_0\mathcal{B}_\epsilon^+$.

Let $A = \sum_{\|\alpha\| \leq r} a_\alpha(x) D^\alpha$ where the $a_\alpha(x)$ are defined and smooth in the box $\mathcal{B}_{\epsilon_0}$ for $\epsilon_0 > 0$. We say A is semi-elliptic at 0 if the constant coefficient heterogeneous operator $A_0 = \sum_{\|\alpha\| = r} a_\alpha(0) D^\alpha$ is semi-elliptic. Note that if A is elliptic at 0 then it is also elliptic at every point close enough to 0.

<u>Theorem</u>. For small enough $\epsilon > 0$ we can find an operator

$$A^{\#} = \sum_{\|\alpha\| \leq r} a^{\#}_\alpha(x)\, D^\alpha$$

with $a^{\#}_\alpha(x)$ defined and smooth on X such that $A^{\#} = A$ inside $\mathcal{B}_\epsilon$ and $A^{\#} = A_0$ outside $\mathcal{B}_{2\epsilon}$; and moreover we have an exact square for $-\infty < k < n < \infty$

$$\begin{array}{ccc} L^p_n(X) & \xrightarrow{A^{\#}} & L^p_{n-r}(X) \\ \downarrow i & & \downarrow i \\ L^p_k(X) & \xrightarrow{A^{\#}} & L^p_{k-r}(X) \end{array} \quad .$$

<u>Proof</u>. We introduce a change of coordinates $x_i = \lambda^{\sigma/\sigma_i} \tilde{x}_i$. Then $\tilde{D}^\alpha = \lambda^{\|\alpha\|} D^\alpha$. The operator A transforms into an operator

$$\tilde{A}_\lambda = \sum_{\|\alpha\| \leq r} \lambda^{r - \|\alpha\|} a_\alpha(\lambda^{\sigma/\sigma_1} \tilde{x}_1, \ldots, \lambda^{\sigma/\sigma_n} \tilde{x}_n)\, \tilde{D}^\alpha$$

defined on $\|x\| \leq \epsilon/\lambda$. Take $\lambda \leq \epsilon_0/2$, let $\psi(x)$ be a smooth function equal to 1 inside $\mathcal{B}_1$ and 0 outside $\mathcal{B}_2$, and put $\tilde{A}_0 = \sum_{\|\alpha\| = r} a_\alpha(\tilde{x}) \tilde{D}^\alpha$ and

$$\tilde{A}^*_\lambda = \lambda^2 \varphi(x) \tilde{A}_\lambda + [1 - \varphi(x)] \tilde{A}_0$$

Then $\tilde{A}^*_\lambda \to \tilde{A}_0$ as $\lambda \to 0$ in the sense that the coefficients are equal outside $\mathcal{B}_2$ and converge uniformly in C^∞ inside $\mathcal{B}_1$. Therefore the square for the operator $\tilde{A}^*_\lambda$ will be exact for small enough λ. Then the same is true for $\tilde{A}^{\#}_\lambda = \lambda^{-r} \tilde{A}^*_\lambda$, and for its transform $A^{\#}_\lambda$ back to the x coordinates from the $\tilde{x}$ coordinates.

But $A^{\#}_{\lambda} = A$ inside $\mathcal{B}_{\epsilon}$ for $\epsilon = \epsilon_0/\lambda$ and $A^{\#}_{\lambda} = A_0$ outside $\mathcal{B}_{2\epsilon}$. This completes the proof.

An analogous result holds for boundary value problems. Let $A = \sum_{\|(\alpha,\beta)\| \leq r} a_{\alpha\beta}(x,y) D_x^{\alpha} D_y^{\beta}$ and $B^j = \sum_{\|(\alpha,\beta)\| \leq r_j} b^j_{\alpha\beta}(x,y) D_x^{\alpha} D_y^{\beta}$ be operators with smooth coefficients in the half-box $\mathcal{B}^+_{\epsilon_0}$. Put

$$A_0 = \sum_{\|(\alpha,\beta)\| = r} a_{\alpha\beta}(0,0)\, D_x^{\alpha}\, D_y^{\beta}$$

$$B_0^j = \sum_{\|(\alpha,\beta)\| = r_j} b^j_{\alpha\beta}(0,0)\, D_x^{\alpha}\, D_y^{\beta} .$$

We say A_0 is semi-elliptic at 0 and the B^j are complementing boundary conditions at 0 if the same is true of A_0 and B_0. Note that if this condition holds at 0 then it holds also at every point in the boundary $X \times \{0\}$ close enough to 0, since the condition says that a certain determinant depending continuously on the $a_{\alpha\beta}$ and $b^j_{\alpha\beta}$ does not vanish on the set $\xi \neq 0$; and by heterogeneity it is enough if this holds for $\|\xi\| = 1$, which is compact.

<u>Theorem</u>. For small enough $\epsilon > 0$ there exist operators $A^{\#}$ and $B^{j\#}$ with smooth coefficients on $X \times Y^+$ agreeing with A and B^j on $\mathcal{B}^+_{\epsilon}$, and agreeing with A_0 and B_0^j outside $\mathcal{B}^+_{2\epsilon}$, such that we have an exact square for $\sigma/\rho p + \max r_j < k < n < \infty$

$$\begin{array}{ccc} L^p_n(X \times Y^+) & \xrightarrow{\mathcal{C}^{\#}} & L^p_{n-r}(X \times Y^+) \overset{m}{\underset{j=1}{\oplus}} \partial L^p_{n-r_j}(X) \\ \downarrow & & \downarrow \\ L^p_k(X \times Y^+) & \xrightarrow{\mathcal{C}^{\#}} & L^p_{k-r}(X \times Y^+) \overset{m}{\underset{j=1}{\oplus}} \partial L^p_{k-r_j}(X) \end{array}$$

where $\mathcal{C}^{\#}(f) = (A^{\#}f,\ B^{1\#}f, \ldots, B^{m\#}f)$.

11. Let A be an operator of weight r with smooth coefficients semi-elliptic at 0.

<u>Theorem.</u> For small enough $0 < \delta < \epsilon$ and for $-\infty < k < n < \infty$ with $k \geq n-1$.

(1) $\|f\|_{L_n^p(\mathcal{B}_\delta)} \leq C(\|Af\|_{L_{n-r}^p(\mathcal{B}_\epsilon)} + \|f\|_{L_k^p(\mathcal{B}_\epsilon)})$ for all $f \in L_n^p(\mathcal{B}_\epsilon)$.

(2) If $f \in L_k^p(\mathcal{B}_\epsilon)$ and $Af \in L_{n-r}^p(\mathcal{B}_\epsilon)$ then $f \in L_n^p(\mathcal{B}_\delta)$.

(3) If $g \in L_{k-r}^p(\mathcal{B}_\delta/\partial)$ then there exists $f \in L_k^p(\mathcal{B}_\epsilon/\partial)$ with $g - Af \in L_{n-r}^p(\mathcal{B}_\epsilon/\partial)$.

<u>Proof.</u> Choose an operator $A^{\#}$ as in Section 10 equal to A on $\mathcal{B}_\epsilon$ and with an exact square. Let $\psi(x)$ be a smooth function equal to 1 on $\mathcal{B}_\delta$ and 0 outside $\mathcal{B}_\epsilon$. Then the commutator $[A,\psi]$ is a partial differential operator of weight at most $n-1 \leq k$.

(1) If $f \in L_n^p(\mathcal{B}_\epsilon)$ then $\psi f \in L_n^p(\mathcal{B}_\epsilon/\partial)$ so

$$\|f\|_{L_n^p(\mathcal{B}_\delta)} \leq \|\psi f\|_{L_n^p(X)} \leq C(\|A^{\#}\psi f\|_{L_{n-r}^p(X)} + \|\psi f\|_{L_k^p(X)})$$

$$\leq C(\|\psi A^{\#} f\|_{L_{n-r}^p(X)} + \|[\psi,A^{\#}]f\|_{L_{n-r}^p(X)} + \|\psi f\|_{L_k^p(X)})$$

$$\leq C(\|Af\|_{L_{n-r}^p(\mathcal{B}_\epsilon)} + \|f\|_{L_k^p(\mathcal{B}_\epsilon)}).$$

(2) Suppose $f \in L_k^p(\mathcal{B}_\epsilon)$ and $Af \in L_{n-r}^p(\mathcal{B}_\epsilon)$. Then $\psi f \in L_k^p(X)$. Also

$$A^{\#}\psi f = [A^{\#},\psi]f + \psi A^{\#} f$$

and $\psi A^{\#} f = \psi Af \in L_{n-r}^p(X)$ while $[A^{\#},\psi]f \in L_{n-r}^p(X)$ as well. Thus $A^{\#}\psi f \in L_{n-r}^p(X)$. By the exactness of the square for $A^{\#}$ we have $\psi f \in L_n^p(X)$ so $f \in L_n^p(\mathcal{B}_\delta)$.

(3) If $g \in L_{k-r}^p(\mathcal{B}_\delta/\partial)$ we can find $f \in L_k^p(X)$ with $g - A^{\#} f \in L_{n-r}^p(X)$. Then $g - A^{\#}(\psi f) = g - \psi A^{\#} f + [\psi,A^{\#}]f$ $= \psi(g - A^{\#} f) + [\psi,A^{\#}]f \in L_{n-r}^p(X)$ also. Then if $\tilde{f} = \psi f$ we have $\tilde{f} \in L_k^p(\mathcal{B}_\epsilon/\partial)$ and $g - A^{\#}\tilde{f} \in L_{n-r}^p(\mathcal{B}_\epsilon/\partial)$. This completes the proof.

In the same way we prove the following:

Theorem. Let A be an operator of weight r with smooth coefficients semi-elliptic at 0 and let B^j $(1 \leq j \leq m)$ be smooth boundary operators with weights r_j satisfying the complementing boundary conditions at 0. Then for small enough $0 < \delta < \epsilon$ and for $\sigma/\rho p + \max r_j < k < n < \infty$

(1) $$\|f\|_{L^p_n(\mathcal{B}^+_\delta)} \leq C\left(\|Af\|_{L^p_{n-r}(\mathcal{B}^+_\epsilon)} + \sum_{j=1}^{m} \|B^j f\|_{\partial L^p_{n-r_j}(\partial_0\mathcal{B}^+_\epsilon)} + \|f\|_{L^p_k(\mathcal{B}^+_\epsilon)}\right)$$

for all $f \in L^p_n(\mathcal{B}^+_\epsilon)$.

(2) If $f \in L^p_k(\mathcal{B}^+_\epsilon)$ and $Af \in L^p_{n-r}(\mathcal{B}^+_\epsilon)$ and $B^j f \in \partial L^p_{n-r_j}(\partial_0\mathcal{B}^+_\epsilon)$ then $f \in L^p_n(\mathcal{B}^+_\epsilon)$.

(3) If $g \in L^p_{k-r}(\mathcal{B}^+_\delta/\partial_e)$ and $h_j \in \partial L^p_{k-r_j}(\partial_0\mathcal{B}^+_\delta/\partial)$ then there exists $f \in L^p_k(\mathcal{B}^+_\epsilon/\partial_e)$ with $g - Af \in L^p_{n-r}(\mathcal{B}^+_\epsilon/\partial_e)$ and $h_j - B^j f \in \partial L^p_{n-r_j}(\partial_0\mathcal{B}^+_\epsilon/\partial)$.

12. Let M be a compact foliated manifold with boundary (but no corners) whose boundary is compatible with the foliation. Let A be a semi-elliptic operator on M, which means that the representation of A in each coordinate chart is semi-elliptic at each point. Let $B^j (1 \leq j \leq m)$ be complementing boundary conditions for A, which also means the same is true for the representations of the B^j in each coordinate chart at each point. Let A have weight r and B^j have weight r_j.

Theorem. If $\sigma/\rho p + \max r_j < k < n < \infty$ we have an exact square

$$\begin{array}{ccc} L^p_n(M) & \xrightarrow{\mathcal{C}} & L^p_{n-r}(M) \overset{m}{\underset{j=1}{\oplus}} \partial L^p_{n-r_j}(\partial M) \\ \downarrow & & \downarrow \\ L^p_k(M) & \xrightarrow{\mathcal{C}} & L^p_{k-r}(M) \overset{m}{\underset{j=1}{\oplus}} \partial L^p_{k-r_j}(\partial M) \end{array}$$

where $\mathcal{C}f = (Af, B^1f,\dots,B^mf)$.

Proof. Since M is compact we may cover M with charts $\varphi : U \subseteq X \to M$ or $\varphi : U \subseteq X \times Y^+ \to M$ so that each U contains $\mathcal{B}_\delta \subseteq \mathcal{B}_\epsilon \subseteq U$ or $\mathcal{B}^+_\delta \subseteq \mathcal{B}^+_\epsilon \subseteq U$ as before and the $\varphi(\mathcal{B}_{\delta/2})$ and $\varphi(\mathcal{B}^+_{\delta/2})$ cover U. Then we will also have a partition of unity consisting of smooth ψ with $\psi = 0$ outside $\mathcal{B}_\delta$ or $\mathcal{B}^+_\delta$ and $\Sigma\psi = 1$.

(1) If $f \in L^p_n(M)$ we have

$$\|f\|_{L^p_n(M)} \le C\left(\sum \|f\|_{L^p_n(\mathcal{B}_\delta)} + \sum \|f\|_{L^p_n(\mathcal{B}^+_\delta)}\right)$$

$$\le C\left(\sum \|Af\|_{L^p_{n-r}(\mathcal{B}_\epsilon)} + \sum \|f\|_{L^p_k(\mathcal{B}_\epsilon)} + \sum \|Af\|_{L^p_{n-r}(\mathcal{B}^+_\epsilon)} + \sum \|B^jf\|_{\partial L^p_{n-r_j}(\partial_0\mathcal{B}^+_\epsilon)} + \sum \|f\|_{L^p_k(\mathcal{B}^+_\epsilon)}\right).$$

Then

$$\|f\|_{L^p_n(M)} \le C\left(\|Af\|_{L^p_{n-r}(M)} + \sum \|B^jf\|_{\partial L^p_{n-r_j}(\partial M)} + \|f\|_{L^p_k(M)}\right)$$

which is Gårding's inequality and proves the first map in the sequence is one-to-one with closed range.

(2) If $f \in L^p_k(M)$ and $Af \in L^p_{n-r}(M)$ and $B^j f \in \partial L^p_{n-r_j}(\partial M)$ then the same holds on each $\mathcal{B}_\epsilon$ or $\mathcal{B}^+_\epsilon$. Then $f \in L^p_n$ on each $\mathcal{B}_\delta$ or $\mathcal{B}^+_\delta$ so $f \in L^p_n(M)$.

(3) If $g \in L^p_{k-r}(M)$ and $h_j \in \partial L^p_{k-r_j}(\partial M)$ then for the partition of unity used before $\psi g \in L^p_{k-r}(\mathcal{B}_\delta/\partial)$ or $\psi g \in L^p_{k-r}(\mathcal{B}^+_\delta/\partial_e)$ and $\psi h_j \in \partial L^p_{k-r_j}(\partial_0 \mathcal{B}^+_\delta/\partial)$. Then we can find $\tilde{f}$ in $L^p_k(\mathcal{B}_\epsilon/\partial)$ or $L^p_k(\mathcal{B}^+_\epsilon/\partial_e)$ with $\psi g - A\tilde{f} \in L^p_{n-r}(\mathcal{B}_\epsilon/\partial)$ or $L^p_{n-r}(\mathcal{B}^+_\epsilon/\partial_e)$ and $\psi h_j - B^j\tilde{f} \in \partial L^p_{n-r_j}(\partial_0 \mathcal{B}^+_\epsilon/\partial)$. Let $f = \Sigma \tilde{f}$. Then
$g - Af = \Sigma(\psi g - A\tilde{f}) \in L^p_{n-r}(M)$ and $h - B^j f = \Sigma\, \psi h_j - B^j \tilde{f} \in \partial L^p_{n-r_j}(\partial M)$.
This completes the proof.

13. <u>Theorem</u>. Let

$$\begin{array}{ccc} E & \xrightarrow{\ \ell\ } & F \\ {\scriptstyle m}\downarrow & & \downarrow{\scriptstyle p} \\ G & \xrightarrow[\ q\ ]{} & H \end{array}$$

be an exact square. If the maps m and p are compact then ℓ and q have isomorphic finite-dimensional kernels and cokernels.

<u>Proof</u>. Let K_ℓ and K_q be the kernels of ℓ and q respectively. Then there is a natural map $m: K_\ell \to K_q$. If $x \in K_\ell$ and $mx = 0$ then $\ell x = 0$ and $mx = 0$ so $x = 0$. Thus $m: K_\ell \to K_q$ is one-to-one. Also if $y \in K_q$ then $qy = 0 = p0$ so there exists $x \in E$ and $mx = y$ and $\ell x = 0$. Then $x \in K_\ell$ so $m: K_\ell \to K_q$ is onto as well. Hence $m: K_\ell \to K_q$ is an isomorphism. But it is also compact. It is well known that any locally compact Banach space is finite dimensional. Therefore K_ℓ and K_q are finite dimensional.

Write $E' = E/\text{Ker}\,\ell$ and $G' = G/\text{Ker}\,q$. Then there is a commutative diagram

$$\begin{array}{ccc} E' & \xrightarrow{\ \ell'\ } & F \\ {\scriptstyle m'}\big\downarrow & & \big\downarrow{\scriptstyle p} \\ G' & \xrightarrow[\ q'\ ]{} & H \end{array}$$

which we call the reduced square. We can easily check that the reduced square is also exact.

<u>Lemma</u>. The image of ℓ' is closed in F.

<u>Proof</u>. Since ℓ' is one-to-one this is equivalent to asserting that ℓ' is an isomorphism onto its range. If not then we can find a sequence $x_n \in E'$ with $\|x_n\| = 1$ and $\ell' x_n \to 0$. Since m' is compact we can find a subsequence which we also write as x_n with $m' x_n$ convergent. Then $\ell' x_n$ and $m' x_n$ are Cauchy. But $\ell' \oplus m' : E' \to F \oplus G'$ is one-to-one with closed range since the reduced square is exact. Therefore x_n is Cauchy also, so $x_n \to x$. Then $\|x\| = 1$. But $\ell' x_n \to \ell' x$ and $\ell' x_n \to 0$ so $\ell' x = 0$. Since ℓ' is one-to-one, $x = 0$. This is a contradiction. Therefore ℓ' has closed range.

We can now finish the proof. Let $F'' = F/\ell'(E')$ and $H'' = H/\overline{q'(G')}$ where we need to take the closure $\overline{q'(G')}$ to guarantee H'' is a Banach space. Then there is an induced map $p'' : F'' \to H''$. Clearly p'' is continuous, compact, and onto. For if $x/\overline{q'(G')}$ is an element of H'' we can write $x = p(y) + q(z)$ and then $p''(y/\ell'(E')) = p(y)/\overline{q'(G')} = x/\overline{q(G')}$. We claim that p'' is also one-to-one. For let $x/\ell'(E')$ be a non-zero element of F''. By the Hahn-Banach theorem we can find a continuous linear functional λ on F with $\lambda | \ell'(E') = 0$ and $\lambda(x) = 1$. The exact sequence

$$0 \to E' \to F \oplus G' \to H \to 0$$

gives rise to a dual exact sequence

$$0 \to H^* \to F^* \oplus G'^* \to E'^* \to 0.$$

Let $\tilde{\lambda}$ be the linear functional on $F^* \oplus G'^*$ equal to λ on F^* and zero on G'^*. Then $\tilde{\lambda} \to 0$ in E'^* so we can find a linear functional μ on H with $\mu \cdot p = \lambda$ on F and $\mu \cdot q' = 0$ on G'. Then $\mu(p(x)) = \lambda(x) = 1$ but $\mu(\overline{q'(G')}) = 0$ so $p(x) \notin \overline{q'(G')}$. Thus p'' is one-to-one.

Thus p'' is an isomorphism and compact. Therefore F'' and H'' are finite dimensional. To complete the proof we can form a second reduced square and $F' = \ell'(E')$ and $H' = \overline{q'(G')}$. Then it is easy to check that this square

$$\begin{array}{ccc} E' & \longrightarrow & F' \\ \downarrow & & \downarrow \\ G' & \longrightarrow & H' \end{array}$$

is also exact. Moreover the top map $E' \to F'$ is now an isomorphism. It follows that the bottom map is also an isomorphism. This shows that q also has a closed range.

14. As a corollary we have the following:

<u>Theorem</u>. Let M be a compact foliated manifold with boundary. Let A be a semi-elliptic operator of weight r and let B^j be complementing boundary conditions of weights r_j. If $1 < p < \infty$ and

$$n > \sigma/\rho p + \max r_j$$

then the map

$$L^p_n(M) \xrightarrow{\ \mathcal{C}\ } L^p_{n-r}(M) \ \bigoplus_{j=1}^{m} \partial L^p_{n-r_j}(\partial M)$$

given by $\mathcal{C}f = (Af, B^1 f, \ldots, B^m f)$ has a finite dimensional kernel and its range is closed with finite codimension; moreover the kernel and cokernel are independent of n.

It follows that for $n > \sigma/\rho p + \max r_j$ the kernel of $\mathcal{C}$ is the finite dimensional space of smooth functions $f \in \mathcal{C}_\infty(M)$ with $Af = 0$ on M and $B^j f = 0$ on ∂M. To obtain the proper corresponding smoothness assertion for the cokernel we need to do a little more work. The cokernel of $\mathcal{C}$ will be expressed by a finite number of linear relations of the form

$$\langle g,\gamma\rangle_M + \sum_{j=1}^{m} \langle h_j,\eta_j\rangle_{\partial M} = 0$$

with γ a linear functional on $L^p_{n-r}(M)$ and η_j a linear functional on $\partial L^p_{n-r_j}(\partial M)$.

Theorem. If $\max r_j < r$ then the cokernel of $\mathcal{C}$ can be expressed by a finite number of linear relations

$$\langle g,\gamma\rangle_M + \sum_{j=1}^{m} \langle h_j,\eta_j\rangle_{\partial M} = 0$$

with $\gamma \in \mathcal{C}_\infty(M)$ and $\eta_j \in \mathcal{C}_\infty(\partial M)$. If $\max r_j - r = k \geq 0$ then for $g \in L^p_{n-r}(M)$ we have the ith normal derivative $\frac{\partial^i g}{\partial \nu^i}$ well defined for $\sigma i/\rho \leq k$. In this case the cokernel of $\mathcal{C}$ can be expressed by a finite number of linear relations

$$\langle g,\gamma\rangle_M + \sum_{\sigma i/\rho \leq k} \langle \frac{\partial^i g}{\partial \nu^i}, \chi_i\rangle_{\partial M} + \sum_{j=1}^{m} \langle h_j,\eta_j\rangle_{\partial M} = 0$$

with $\gamma \in \mathcal{C}_\infty(M)$, $\chi_i \in \mathcal{C}_\infty(\partial M)$ and $\eta_j \in \mathcal{C}_\infty(\partial M)$.

If $g \in \mathcal{C}_\infty(M)$ and $h_j \in \mathcal{C}_\infty(\partial M)$ satisfy these relations then there exists an $f \in \mathcal{C}_\infty(M)$ with $Af = g$ and $B^j f = h_j$, and conversely.

15. In order to prove these results we introduce bigraded spaces

$$L^p_{n,m}(X \times Y) = \{f \in \mathcal{S}(X \times Y)^*:\ W(D_x, D_y)^n W(D_x)^m f \in L^p(X \times Y)\}$$

where $W(\xi,\eta) = (1+\xi_1^{2\sigma_1}+\dots+\xi_n^{2\sigma_n} + \eta^{2\rho})^{1/2\sigma}$ and

$W(\xi) = (1+\xi_1^{2\sigma_1}+\dots+\xi_n^{2\sigma_n})^{1/2\sigma}$. The norm

$$\|f\|_{L^p_{n,m}(X\times Y)} = \|W(D_x, D_y)^n W(D_x)^m f\|_{L^p(X\times Y)}$$

measures derivatives of weight $n+m$ in the x-directions but only derivatives of weight n in the y-directions. There are corresponding spaces $L^p_{n,m}(X\times Y^+)$, $L^p_{n,m}(X\times Y^+/0)$, and $\partial L^p_{n,m}(X)$ for $n > \sigma/\rho p$. Then if $A^\#$ and $B^{j\#}$ are operators with almost constant coefficients we have an exact square for $n \geq k > \sigma/\rho p + \max r_j$ and any $m \leq \ell$

$$\begin{array}{ccc}
L^p_{n,-m}(X\times Y^+) & \xrightarrow{\ \mathcal{C}\ } & L^p_{n-r,-m}(X\times Y^+) \overset{m}{\underset{j=1}{\oplus}} \partial L^p_{n-r_j,-m}(X) \\
\downarrow & & \downarrow \\
L^p_{k,-\ell}(X\times Y^+) & \xrightarrow{\ \mathcal{C}\ } & L^p_{k-r,-\ell}(X\times Y^+) \overset{m}{\underset{j=1}{\oplus}} \partial L^p_{k-r_j,-\ell}(X)\ .
\end{array}$$

The dual space of $L^p_{n,m}(X\times Y^+)$ is $L^q_{-n,-m}(X\times Y^+/0)$ with $1/p + 1/q = 1$. The dual space of $\partial L^p_{n,m}(X)$ we denote $\partial^* L^q_{-n,-m}(X)$. We know that if $s > n + \sigma/\rho p$ there is an inclusion

$$L^p_s(X) \to \partial L^p_n(X).$$

Since $W(D_x)$ commutes with restriction to the boundary there is an inclusion

$$L^p_{s-m}(X) \to \partial L^p_{n,-m}(X)$$

and by duality an inclusion

$$\partial^* L^p_{-n,m}(X) \to L^p_{m-s}(X).$$

Thus elements of $\partial^* L^p_{-n,m}$ are as smooth as we like for large enough m.

Lemma. If the square

$$\begin{array}{ccc} E & \xrightarrow{\ \ell\ } & F \\ {\scriptstyle m}\downarrow & & \downarrow{\scriptstyle p} \\ G & \xrightarrow[\ q\ ]{} & H \end{array}$$

is exact so is the dual square

$$\begin{array}{ccc} H^* & \xrightarrow{\ q^*\ } & G^* \\ {\scriptstyle p^*}\downarrow & & \downarrow{\scriptstyle m^*} \\ F^* & \xrightarrow[\ \ell^*\ ]{} & E^* \end{array} .$$

Proof. The sequence

$$0 \longrightarrow E \xrightarrow{\ \ell\oplus m\ } F \oplus G \xrightarrow{\ p\ominus q\ } H \longrightarrow 0$$

is exact. Since $(x,y) \to (x,-y)$ is an isomorphism the sequence

$$0 \longrightarrow E \xrightarrow{\ \ell\ominus m\ } F \oplus G \xrightarrow{\ p\oplus q\ } H \longrightarrow 0$$

is also exact. Then so is its dual

$$0 \longrightarrow H^* \xrightarrow{\ p^*\oplus q^*\ } F^* \oplus G^* \xrightarrow{\ \ell^*\ominus m^*\ } E^* \longrightarrow 0.$$

We have a dual exact square for $m \leq \ell$ and $k > \sigma/\rho p + \max r_j$

$$\begin{array}{ccc} L^q_{r-k,\ell}(X\times Y^+/0) \ \bigoplus_{j=1}^{m} \partial^* L^q_{r_j-k,\ell}(X) & \xrightarrow{\ \mathcal{C}^*\ } & L^q_{-k,\ell}(X) \\ \downarrow & & \downarrow \\ L^q_{r-k,n}(X\times Y^+/0) \ \bigoplus_{j=1}^{m} \partial^* L^q_{r_j-k,n}(X) & \xrightarrow{\ \mathcal{C}^*\ } & L^q_{-k,n}(X) \end{array}$$

We can define spaces on boxes as well. Then we prove the following lemma:

Lemma. Suppose $n \geq \ell-1$ and $\gamma \in L^q_{r-k,n}(\mathcal{B}^+_\epsilon/\partial_0)$ and $\eta_j \in \partial^* L^q_{r_j-k,n}(\partial_0 \mathcal{B}^+_\epsilon)$ and $\mathcal{C}^*(\gamma,\eta_1,\ldots,\eta_m) \in L^q_{-k,\ell}(\mathcal{B}^+_\epsilon/\partial_0)$. Then $\gamma \in L^q_{r-k,\ell}(\mathcal{B}^+_\delta/\partial_0)$ and $\eta_j \in \partial^* L^q_{r_j-k,\ell}(\partial_0\mathcal{B}^+_\delta)$.

Proof. If ψ is a smooth function equal to 1 inside $\mathcal{B}^+_\delta$ and zero outside $\mathcal{B}^+_\epsilon$ then

$$\mathcal{C}^*(\psi\gamma,\psi\eta_1,\ldots,\psi\eta_m)f = \langle Af,\psi\gamma\rangle + \Sigma\langle B^jf,\psi\eta\rangle = \langle\psi Af,\gamma\rangle + \Sigma\langle\psi B^jf,\eta\rangle$$

$$= \langle A\psi f,\gamma\rangle + \Sigma\langle B^j\psi f,\eta\rangle + \langle[A,\psi]f,\gamma\rangle + \Sigma\langle[B^j,\psi]f,\eta\rangle$$

$$= \mathcal{C}^*(\gamma,\eta_1,\ldots,\eta_m)\psi f + \langle[A,\psi]f,\gamma\rangle + \Sigma\langle[B^j,\psi]f,\eta\rangle$$

$$\|\mathcal{C}^*(\psi\gamma,\psi\eta)\|_{L^q_{-k,\ell}} = \sup\{|\mathcal{C}^*(\psi\gamma,\psi\eta)f| : \|f\|_{L^p_{k,-\ell}} \leq 1\}.$$

But if $f \in L^p_{k,-\ell}$ then $[A,\psi]f \in L^p_{k-r,\ell+1}$ and $[B^j,\psi]f \in L^p_{k-r_j,\ell+1}$. Then

$$|\langle[A,\psi]f,\gamma\rangle| \leq \|f\|_{L^p_{k,-\ell}} \|\gamma\|_{L^q_{r-k,\ell-1}}$$

$$|\langle[B_j,\psi]f,\eta\rangle| \leq \|f\|_{L^p_{k,-\ell}} \|\eta\|_{\partial^* L^q_{r-k_j,\ell-1}}$$

$$\|\mathcal{C}^*(\psi\gamma,\psi\eta)\|_{L^q_{-k,\ell}} \leq c\|\mathcal{C}^*(\gamma,\eta)\|_{L^q_{-k,\ell}}$$

$$+ c\|\gamma\|_{L^q_{r-k,m}} + c\sum_{j=1}^{m} \|\eta_j\|_{L^q_{r_j-k,m}}$$

Therefore $\mathcal{C}^*(\psi\gamma,\psi\eta) \in L^q_{-k,\ell}(X \times Y^+)$ also. Thus $\psi\gamma \in L^q_{r-k,\ell}(X \times Y^+)$ and $\psi\eta \in \partial^* L^q_{r_j-k,\ell}(X)$, by the exact square. Hence

$$\gamma \in L^q_{r-k,\ell}(\mathcal{B}^+_\delta/\partial_0) \quad\text{and}\quad \eta \in \partial^* L^q_{r_j-k,\ell}(\partial_0\mathcal{B}^+_\delta).$$

Suppose then that $\gamma \in \mathcal{C}_\infty(M)^*$ and $\eta_j \in \mathcal{C}_\infty(\partial M)^*$ form an element of the cokernel of $\mathcal{C}$, in that $\mathcal{C}^*(\gamma,\eta) = 0$. Then for any $k > \sigma/\rho p + \max r_j$ and any m we have locally

$$\gamma \in L^q_{r-k,m}(\mathcal{B}^+_\delta/\partial_0)$$

$$\eta_j \in \partial^* L^q_{r_j-k,m}(\mathcal{B}^+_\delta).$$

This shows η_j is smooth. For γ we know that its image $Z^*C^*\gamma$ in $L^q_{r-k,m}(\mathcal{B}^+_\delta)$ satisfies $A^*(Z^*C^*\gamma) = 0$ where A^* is the operator adjoint to A. Thus we can write normal derivatives of $Z^*C^*\gamma$ in terms of tangential derivatives. Therefore $Z^*C^*\gamma$ is smooth. If $r > \max r_j$ then for large enough p we can choose $k = r$. Then Z^*C^* is an isomorphism and γ itself is smooth. Otherwise the kernel of Z^*C^* is non-trivial: In fact there is an exact sequence

$$0 \to L^p_{n,-m}(\mathcal{B}^+/\partial) \to L^p_{n,-m}(\mathcal{B}^+/\partial_e) \to \bigoplus_j \partial L^p_{n-j,-m}(\partial_0\mathcal{B}^+/\partial) \to 0$$

with the last term representing the normal derivatives which are well-defined; and a dual sequence

$$0 \to \bigoplus_j \partial^* L^q_{j-n,m}(\partial_0\mathcal{B}^+) \to L^q_{-n,m}(\mathcal{B}^+/\partial_0) \xrightarrow{Z^*C^*} L^q_{-n,m}(\mathcal{B}^+) \to 0$$

with the first term representing distributions concentrated on the boundary, looking like a function on the boundary times the normal derivative of a δ-function at the boundary. These provide the extra terms in the linear relations when $r \leq \max r_j$. To be precise, γ maps into a smooth function in $L^q_{r-k,m}(\mathcal{B}^+/\partial_0)$. But this smooth function can also be regarded as the image of a smooth function $\tilde{\gamma}$ in $L^q_{r-k,m}(\mathcal{B}^+)$. Then $\gamma - \tilde{\gamma}$ lies in the kernel and so it can be represented by elements in $\bigoplus_j \partial^* L^q_{j+r-k,m}(\partial_0\mathcal{B}^+)$ which are therefore smooth functions on ∂M times normal derivatives of the δ-function at the boundary.

16. The key to an understanding of parabolic equations is a theorem of Payley and Wiener, which says that if $f \in \mathcal{S}(X \times T)$ is smooth and rapidly decreasing then $f(x,t) = 0$ for $t \leq 0$ if and only if the Fourier transform $\hat{f}(\xi,\theta)$ extends to a holomorphic smooth rapidly decreasing function on $\Xi \times \Theta^-$ where Θ^- is the lower half plane $\mathrm{Im}\theta \leq 0$. Let $M(\xi,\theta)$ be a multiplier which satisfies the hypotheses of Stein's theorem:

$$|\xi^\alpha \theta^\beta D_\xi^\alpha D_\theta^\beta M(\xi,\theta)| \leq C$$

for primitive α and β. Note that if $M(\xi,\theta)$ grows no faster than a polynomial in $\Xi \times \Theta^-$ and is bounded on $\Xi \times \Theta^0$ where Θ^0 is the real line $\mathrm{Im}\theta = 0$ then the Phragmen-Lindelöf theory tells us that $M(\xi,\theta)$ is bounded on all of $\Xi \times \Theta^-$. It follows that the map $M(D_x, D_t)$ taking $L^p(X \times T)$ into itself also takes $L^p(X \times T^+/0)$ into itself, and hence also takes the quotient $L^p(X \times T^-)$ into itself.

We say that the constant coefficient operator

$$A(D_x,D_t) = \sum_{\|(\alpha,\beta)\| \leq r} a_{\alpha\beta} D_x^\alpha D_t^\beta$$

is parabolic if the associated polynomial

$$A(\xi,\theta) = \sum_{\|(\alpha,\beta)\| = r} a_{\alpha\beta}\, \xi^\alpha \theta^\beta$$

has no roots in $\mathrm{Im}\theta \leq 0$ and ξ real except $\xi = \theta = 0$. In this case the operator $G(D_x,D_t)$ defined by

$$G(\xi,\theta) = [1 - \psi(\xi,\theta)]/A(\xi,\theta)$$

which forms the approximate inverse to A will also map $L^p_{n-r}(X \times T^+/0) \to L^p_{n-r}(X \times T^+/0)$ or $L^p_{n-r}(X \times T^-) \to L^p_{n-r}(X \times T^-)$ provided that we are careful to choose $\psi(\xi,\theta) = \psi(\xi)\,\hat{\chi}(\theta)$ so that

$\hat{\chi}(\theta)$ not only has the same Taylor series as 1 at the origin and is smooth and rapidly decreasing at ∞, but also so that $\hat{\chi}(\theta)$ is holomorphic in $\operatorname{Im}\theta \leq 0$. But the function $\hat{\chi}$ of Section 5 has just these properties since $\chi(t) = 0$ for $t \leq 1$.

Likewise we say that B^j are a set of parabolic complementing boundary conditions along the boundary $y = 0$ if when we write

$$B^j(D_x, D_y, D_t) = \sum_{\|(\alpha,\beta,\gamma)\| \leq r_j} b^j_{\alpha\beta\gamma} D_x^{\alpha} D_y^{\beta} D_t^{\gamma}$$

then the associated polynomials

$$B^j(\xi,\eta,\theta) = \sum_{\|(\alpha,\beta,\gamma)\| = r_j} b^j_{\alpha\beta\gamma} \xi^{\alpha}\eta^{\beta}\theta^{\gamma}$$

are linearly independent modulo $A^+(\xi,\eta,\theta)$ as polynomials in η for all ξ and θ with ξ real and $\operatorname{Im}\theta \leq 0$ except $\xi = \theta = 0$. In this case the approximate inverses $H_j(D_x, y, D_t)$ can be chosen so as to map

$$H_j : \partial L^p_{n-r_j}(X \times T^+/0) \to L^p_n(X \times Y^+ \times T^+/0)$$

and also

$$H_j : \partial L^p_{n-r_j}(X \times T^-) \to L^p_n(X \times Y^+ \times T^-).$$

Here the space $\partial L^p_{n-r_j}(X \times T^+/0)$ represents boundary values along $y = 0$ if functions in $L^p_{n-r_j}(X \times Y^+ \times T^+/0)$, and likewise $\partial L^p_{n-r_j}(X \times T^-)$ represents boundary values along $y = 0$ of functions in $L^p_{n-r_j}(X \times Y^+ \times T^-)$. Only two minor modifications of the previous proof are necessary to preserve the support in $t \geq 0$. The first is to replace $\psi(\xi)$ by $\psi(\xi)\hat{\chi}(\theta)$; the second is to take derivatives with $W(D_x)$ and $W'(D_t)$ where $W(\xi)$ is as before and $W'(\theta) = (1 + i\theta)^{\tau/\sigma}$ where τ is the weight of t. This makes $W'(\theta)$ holomorphic in $\operatorname{Im}\theta \leq 0$ also, and so $W'(D_t)$ preserves support in $t \geq 0$. If A and B^j have variable coefficients we require these conditions to hold at each point in order for A to be parabolic and the B^j to be parabolic complementing boundary conditions. Then repeating the previous arguments we obtain exact squares of the sort

$$\begin{array}{ccc} L^p_n(X\times Y^+\times T^+/0) & \to & L^p_{n-r}(X\times Y^+\times T^+/0) \overset{m}{\underset{j=1}{\oplus}} \partial L^p_{n-r_j}(X\times T^+/0) \\ \downarrow & & \downarrow \\ L^p_k(X\times Y^+\times T^+/0) & \to & L^p_{k-r}(X\times Y^+\times T^+/0) \overset{m}{\underset{j=1}{\oplus}} \partial L^p_{k-r_j}(X\times T^+/0) \end{array}$$

and also

$$\begin{array}{ccc} L^p_n(X\times Y^+\times T^-) & \to & L^p_{n-r}(X\times Y^+\times T^-) \overset{m}{\underset{j=1}{\oplus}} \partial L^p_{n-r_j}(X\times T^-) \\ \downarrow & & \downarrow \\ L^p_k(X\times Y^+\times T^-) & \to & L^p_{k-r}(X\times Y^+\times T^-) \overset{m}{\underset{j=1}{\oplus}} \partial L^p_{k-r_j}(X\times T^-). \end{array}$$

To consider problems on manifolds suppose that M is a compact foliated manifold with corners, with a globally defined time function t, defining the foliation. We suppose moreover that ∂M decomposes naturally into three parts, an initial time boundary $\partial_\alpha M$ defined by $\{t = \alpha\}$, a final time boundary $\partial_\omega M$ defined by $\{t = \omega\}$, and a space boundary $\partial_S M$ transversal everywhere to the time foliation.

M :

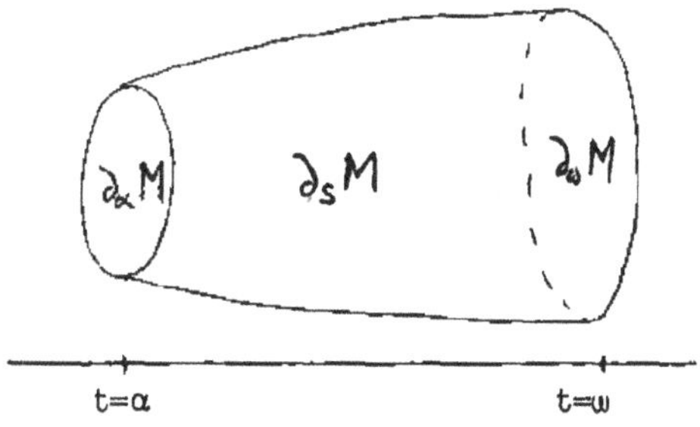

We consider a parabolic operator A on M and parabolic complementing boundary conditions on $\partial_S M$ only. We phrase the initial condition implicitly by restricting our attention to the spaces $L_n^p(M/\partial_\alpha)$ which is the closure of the smooth functions that vanish together with all their derivatives at $t = \alpha$. We can then repeat all of the previous arguments locally using spaces $L_n^p(X)$ in the interior, $L_n^p(X \times T^+/0)$ on $\partial_\alpha M$, $L_n^p(X \times T^-)$ on $\partial_\omega M$, $L_n^p(X \times Y^+)$ on $\partial_S M$, $L_n^p(X \times Y^+ \times T^+/0)$ on $\partial_\alpha M \cap \partial_S M$ and $L_n^p(X \times Y^+ \times T^-)$ on $\partial_\omega M \cap \partial_S M$.

We suppose that the time direction has weight τ and the space directions weight σ in the local charts. An easy algebraic argument shows that σ must be an even multiple of τ. Thus σ is the least common multiple also. Therefore it is no loss to take $\tau = 1$. The only difference from the previous theory is that due to the causal nature of parabolic problems the finite dimensional kernel and cokernel are in fact zero. Thus we have the following:

<u>Theorem.</u> The map $\mathcal{C} f = (Af, B^j f)$ defines an isomorphism

$$\mathcal{C}_\infty(M/\partial_\alpha) \xrightarrow{\mathcal{C}} \mathcal{C}_\infty(M/\partial_\alpha) \bigoplus_{j=1}^{m} \mathcal{C}_\infty(\partial_S M/\partial_\alpha)$$

and also for each $n > 1/p + \max r_j$ an isomorphism

$$L_n^p(M/\partial_\alpha) \to L_{n-r}^p(M/\partial_\alpha) \bigoplus_{j=1}^{m} \partial L_{n-r_j}^p(\partial_S M/\partial_\alpha).$$

17. <u>Proof.</u> We know from the previous arguments that the map

$$L_n^p(M/\partial_\alpha) \to L_{n-r}^p(M/\partial_\alpha) \bigoplus_{j=1}^{m} \partial L_{n-r_j}^p(\partial_S M/\partial_\alpha)$$

has finite dimensional kernel, closed range and finite dimensional cokernel, all independent of n if $n > 1/p + \max r_j$. Moreover, the kernel consists of smooth functions $f \in \mathcal{C}_\infty(M/\partial_\alpha)$ and the cokernel is represented by linear relations

$$\langle g,\gamma\rangle_n + \sum_{i=1}^{k} \langle \frac{\partial^i}{\partial \nu^i} g,\ \chi_i\rangle_{\partial_S M} + \sum_{j=1}^{m} \langle h_i,\eta_i\rangle$$

with $\gamma \in \mathcal{C}_\infty(M/\partial_\omega)$, $\chi_i \in \mathcal{C}_\infty(\partial_S M/\partial_\omega)$ and $\eta_i \in \mathcal{C}_\infty(\partial_S M/\partial_\omega)$.

Suppose then that $\alpha \leq \beta < \gamma \leq \omega$. We can consider $\dim \operatorname{Ker}(\beta,\gamma)$ and $\dim \operatorname{Coker}(\beta,\gamma)$ as functions of β and γ by considering the same problem on the smaller manifold $\tilde{M}$ where $\beta \leq t \leq \gamma$. Then we see that as functions of β or γ, $\dim \operatorname{Ker}(\beta,\gamma)$ is continuous from the left and $\dim \operatorname{Coker}(\beta,\gamma)$ from the right. Moreover if $\operatorname{Ker}(\alpha,\omega)$ is not zero then $\dim \operatorname{Ker}(\beta,\omega)$ must experience a sudden drop for some $\beta < \omega$ as $\beta \to \omega$, while if $\operatorname{Coker}(\alpha,\omega)$ is not zero then $\dim \operatorname{Coker}(\alpha,\gamma)$ must experience a sudden drop for some $\gamma > \alpha$ as $\gamma \to \alpha$.

However, we can also use the ideas of index theory. We can topologize the space of operators A and B with the $\mathcal{C}^\infty$ topology as appropriate sections of vector bundles over M. Then the set of parabolic A and parabolic complementing boundary conditions B is open. Moreover the map

$$\mathcal{C}(A,B):\ L^p_n(M/\partial_\alpha) \to L^p_{n-r}(M/\partial_\alpha) \bigoplus_{j=1}^{m} \partial L^p_{n-r_j}(\partial_S M/\partial_\alpha)$$

depends continuously on A and B and always has finite dimensional kernel, closed range, and finite-dimensional cokernel. Such a map is called a Fredholm map. It is a classical result that if E and F are Banach spaces then the set $\mathcal{F}$ of Fredholm maps is open in $L(E,F)$ and the index $i(\ell) = \dim \operatorname{Ker} \ell - \dim \operatorname{Coker} \ell$ is continuous on $\mathcal{F}$ and hence constant on each component. Thus the index $\dim \operatorname{Ker} \mathcal{C}(A,B) - \dim \operatorname{Coker} \mathcal{C}(A,B)$ remains constant if we vary A and B continuously.

Now our conditions on M guarantee that M is diffeomorphic to a product $M = N \times [\alpha,\omega]$ where N is a manifold with boundary and the time function t is the projection on the second coordinate. But from this it is easy to see that varying the initial and final times β and γ is equivalent (by changing the time scale) to continuously varying the operators A and B on the original M. Thus dim Ker(β,γ) - dim Coker(β,γ) is a constant. Since dim Ker(β,γ) is continuous from the left in β and γ while dim Coker(β,γ) is continuous from the right, it follows that each is constant. Then neither can experience a sudden drop so both must be zero to begin with. Therefore

$$\mathcal{C}: L_n^p(M/\partial_\alpha) \to L_{n-r}^p(M/\partial_\alpha) \bigoplus_{j=1}^{m} \partial L_{n-r_j}^p(\partial_S M/\partial_\alpha)$$

is an isomorphism.

18. We also have Garding's inequality and a regularity theorem for parabolic boundary value problems. Let $M_{\alpha\omega}$ be a manifold with corners foliated by a time function $t: M_{\alpha\omega} \to [\alpha,\omega]$ such that $\partial M_{\alpha\omega}$ consists of three parts; the initial and final boundaries ∂_α and ∂_ω and the space boundary $\partial_S M_{\alpha\omega}$ which is everywhere transversal to the time foliation. If $\alpha \leq \beta < \gamma \leq \omega$ we write $M_{\beta\gamma}$ for the part of $M_{\alpha\omega}$ with $\beta \leq t \leq \gamma$. Then $M_{\beta\gamma}$ is again a foliated manifold with corners. We write $L_n^p[\beta,\gamma]$ for $L_n^p(M_{\beta\gamma})$ and $\partial L_n^p[\beta,\gamma]$ for $\partial L_n^p(\partial_S M_{\beta\gamma})$.

<u>Theorem</u>. Let A be a parabolic operator of weight r and let B^j be parabolic complementing boundary conditions of weights r_j. Suppose $1 < p < \infty$, $n > k > 1/p + \max r_j$, and $\alpha < \pi < \omega$. If $f \in L_k^p[\alpha,\omega]$, $Af \in L_{n-r}^p[\alpha,\omega]$ and $B^j f \in \partial L_{n-r_j}^p[\alpha,\omega]$ then $f \in L_n^p[\pi,\omega]$. Also for any $\ell > -\infty$ we have

$$\|f\|_{L_n^p[\pi,\omega]} \leq C(\|Af\|_{L_{n-r}^p[\alpha,\omega]} + \|B^j f\|_{\partial L_{n-r_j}^p[\alpha,\omega]} + \|f\|_{L_\ell^p[\alpha,\pi]}).$$

Corollary. If $f \in L_k^p[\alpha,\omega]$ for some $k > 1/p + \max r_j$ and satisfies the homogeneous equations $Af = 0$ and $B^j f = 0$ then f is of class $\mathcal{C}^\infty$ for $\alpha < t \le \omega$. Moreover, for $\alpha < \pi < \omega$ and $-\infty < \ell < n < +\infty$ we can find a constant C such that all such solutions satisfy $\|f\|_{L_n^p[\pi,\omega]} \le C\|f\|_{L_\ell^p[\alpha,\pi]}$. Thus an arbitrarily weak estimate in the past gives an arbitrarily strong estimate in the future.

Proof. For the regularity theorem it is enough to prove it when $n \le k+1$; for we can then proceed by induction on n, increasing our degree of regularity by one each time. The proof of the regularity then proceeds in the same way as for semi-elliptic operators, by reducing the problem to a local estimate in boxes and using exact squares. The only difference occurs at the time boundaries $t = \pi$ and $t = \omega$. At $t = \pi$ there is no problem since we are given f in L_k^p back to $t = \alpha < \pi$. At $t = \omega$ there is also no problem, since using the parabolicity we can make all our operators act on spaces $L_n^p(X \times T^-)$ or $L_n^p(X \times Y^+ \times T^-)$.

In the same way we can prove Gårding's inequality

$$\|f\|_{L_n^p[\pi,\omega]} \le C(\|Af\|_{L_{n-r}^p[\alpha,\omega]} + \|B^j f\|_{\partial L_{n-r_j}^p[\alpha,\omega]} + \|f\|_{L_k^p[\alpha,\omega]})$$

with $\|f\|_{L_k^p[\alpha,\omega]}$ in place of $\|f\|_{L_\ell^p[\alpha,\pi]}$. We can improve the one result to the other as follows: If we choose $\tilde{\alpha}$ with $\alpha < \tilde{\alpha} < \pi < \omega$ then $f \in L_n^p[\tilde{\alpha},\omega]$ and we avoid all regularity problems. Translate $t = \pi$ to $t = 0$ and consider the equation in local coordinates (x,y,t) with the origin $(0,0,0)$ corresponding to a point in $\partial_S M$ at $t = \pi$. (The interior estimate is even easier.)

Choose two boxes $\mathcal{B}$ and $\mathcal{B}'$ with

$$\mathcal{B} = \{|x_i| \le \delta_i,\ 0 \le y \le \delta_*,\ 0 \le t \le \delta_\#\}$$

$$\mathcal{B}' = \{|x_i| \le \epsilon_i,\ 0 \le y \le \epsilon_*,\ -\epsilon_\# \le t \le \epsilon_\#\}$$

with $0 < \delta_* < \epsilon_*$, $0 < \delta < \epsilon_\#$ and $0 < \delta_\# < \epsilon_\#$. Introduce $W(D_x, D_y, D_t)$ and $W(D_x, D_t)$ corresponding to $W(\xi, \eta, \theta) = (1 + \xi_1^{2\sigma} + \ldots + \xi_n^{2\sigma} + \eta^{2\sigma} + \theta^2)^{1/2\sigma}$ and $W(\xi, \theta) = (1 + \xi_1^{2\sigma} + \ldots + \xi_n^{2\sigma} + \theta^2)^{1/2\sigma}$. We have bigraded spaces $L^p_{n,m}(X \times Y \times T)$ with norms

$$\| f \|_{L^p_{n,m}(X\times Y\times T)} = \| W(D_x, D_y, D_t)^n W(D_x, D_t)^m f \|_{L^p(X\times Y\times T)}.$$

Then as before we prove that if $n > 1/p + \max r_j$ and $\ell \leq m$ then there is an exact square

$$\begin{array}{ccc} L^p_{n,m}(X\times Y^+\times T) & \to & L^p_{n-r,m}(X\times Y^+\times T) \bigoplus_{j=1}^{m} \partial L^p_{n-r_j,m}(X\times T) \\ \downarrow & & \downarrow \\ L^p_{n,\ell}(X\times Y^+\times T) & \to & L^p_{n-r,\ell}(X\times Y^+\times T) \bigoplus_{j=1}^{m} \partial L^p_{n-r_j,\ell}(X\times T) \end{array}$$

which we prove first for constant coefficient operators and then for a variable coefficient operator close to constant by stability. If $m \leq \ell+1$ we can choose a smooth function ψ equal to 1 inside $\mathcal{B}$ and 0 outside $\mathcal{B}'$, and multiplying by ψ and taking commutators $[A, \psi]$ and $[B^j, \psi]$ we get an estimate

$$\| f \|_{L^p_{n,m}(\mathcal{B})} \leq C(\| Af \|_{L^p_{n-r,m}(\mathcal{B}')} + \sum_{j=1}^{m} \| B^j f \|_{\partial L^p_{n-r_j,m}(\mathcal{B}')} + \| f \|_{L^p_{n,\ell}(\mathcal{B}')}).$$

Such an estimate would still hold if $\mathcal{B}'$ were any smaller box as long as it is larger than $\mathcal{B}$. Therefore in the case where ℓ is much smaller than m we can still obtain the estimate by considering a finite sequence of increasing boxes and using the above estimate repeatedly.

Next we note that using the equation for Af we can estimate y-derivatives of f in terms of x and t derivatives. To be explicit we have a formula

$$\frac{\partial^r}{\partial y^r} f = c \cdot Af + \sum_{\substack{\|(\alpha,\beta,\gamma)\| \leq r \\ \beta < r}} c_{\alpha\beta\gamma} D_x^\alpha D_y^\beta D_t^\gamma f$$

where c and $c_{\alpha\beta\gamma}$ are smooth coefficients.

But

$$\|f\|_{L^p_{n,m}} \leq \sum_{\|(\alpha,\beta,\gamma)\| \leq r} \|D_x^\alpha D_y^\beta D_t^\gamma f\|_{L^p_{k-r,m}}$$

because r must be an integral multiple of σ if A is parabolic. Therefore

$$\|f\|_{L^p_{n,m}(\mathcal{B}')} \leq C(\|g\|_{L^p_{n-r+m}(\mathcal{B}')} + \|f\|_{L^p_{n-1,m+1}(\mathcal{B}')}).$$

Then by induction if $\ell < n$

$$\|f\|_{L^p_{n,\ell-n}(\mathcal{B}')} \leq C(\|g\|_{L^p_{n-r}(\mathcal{B}')} + \|f\|_{L^p_{\ell}(\mathcal{B}')}).$$

We conclude that

$$\|f\|_{L^p_n(\mathcal{B})} \leq C(\|Af\|_{L^p_{n-r}(\mathcal{B}')} + \sum_{j=1}^{m} \|B^j f\|_{\partial L^p_{n-r_j}(\mathcal{B}')} + \|f\|_{L^p_{\ell}(\mathcal{B}')}).$$

If we add together such estimates over $M_{\pi\omega}$ we get

$$\|f\|_{L^p_n[\pi,\omega]} \leq C(\|Af\|_{L^p_{n-r}[\alpha,\omega]} + \sum_{j=1}^{m} \|B^j f\|_{\partial L^p_{n-r_j}[\alpha,\omega]} + \|f\|_{L^p_{\ell}[\alpha,\omega]}).$$

If we choose β and γ with $\alpha < \beta < \gamma < \pi < \omega$ then we can bound $\| f\|_{L^p_n[\beta,\omega]}$ by the same thing as above and

$$\| f\|_{L^p_\ell[\alpha,\omega]} \leq C(\| f\|_{L^p_\ell[\alpha,\pi]} + \| f\|_{L^p_\ell[\gamma,\omega]})$$

so

$$\| f\|_{L^p_n[\beta,\omega]} \leq C(\| Af\|_{L^p_{n-r}[\alpha,\omega]} + \sum_{j=1}^{m} \| B^j f\|_{\partial L^p_{n-r_j}[\alpha,\omega]}$$

$$+ \| f\|_{L^p_\ell[\alpha,\pi]} + \| f\|_{L^p_\ell[\gamma,\omega]}).$$

Now we claim that we can omit the last term. For if not we would find a sequence $f_i \in \mathcal{C}_\infty(M_{\alpha\omega})$ with $Af_i \to 0$ in $L^p_{n-r}[\alpha,\omega]$, $B^j f_i \to 0$ in $\partial L^p_{n-r_j}[\alpha,\omega]$, $f_i \to 0$ in $L^p_\ell[\alpha,\pi]$ but $\| f_i\|_{L^p_\ell[\gamma,\omega]} = 1$. Then by the above estimate $\| f_i\|_{L^p_n[\beta,\omega]} \leq C$. Choose k with $n > k > 1/p + \max r_j$. The inclusion of $L^p_n[\beta,\omega] \to L^p_k[\beta,\omega]$ is compact, so a subsequence f_i would converge in $L^p_k[\beta,\omega]$. Its limit f would satisfy $Af = 0$ and $B^j f = 0$ on $[\beta,\omega]$, $f = 0$ on $[\beta,\pi]$ but $f \neq 0$ on $[\gamma,\omega]$. This contradicts the fact that there are no non-trivial solutions of the homogeneous equation in $L^p_k(M_{\gamma\omega}/\partial_\gamma)$. Therefore we can neglect the last term and the estimate is established.

Part IV: The Heat Equation for Manifolds

1. We consider the Dirichlet problem from Part 1. Let X and Y be compact manifolds with smooth boundaries, and let $h: \partial X \to Y$ be a smooth map on ∂X. We want to find a solution $f: X \to Y$ of $\Delta f = 0$ on X and $f = h$ on ∂X in every relative homotopy class. Let $f_0: X \to Y$ be a representative of the desired relative homotopy class in which we seek a solution, with $f_0 = h$ on ∂X. We shall construct a one-parameter family of deformations $f_t: X \to Y$ for $t \geq 0$ agreeing with f_0 at $t = 0$ and with $f_t = h$ on ∂X for all t, such that f_t converges to a solution f_∞ of $\Delta f_\infty = 0$ on X and $f_\infty = h$ on ∂X. Then f_∞ will be a solution in our given homotopy class.

The maps $f_t: X \to Y$ for $t \geq 0$ come from a map $f: X \times [0,\infty) \to Y$ by setting $f_t(x) = f(x,t)$. We construct f as the solution of the nonlinear parabolic initial boundary value problem

$$(H)\quad \begin{cases} \frac{\partial f}{\partial t} = \Delta f & \text{on } X \times [0,\infty) \\ f(x,0) = f_0(x) & \text{on } X \times 0 \\ f(x,t) = h(x) & \text{on } \partial X \times [0,\infty)\,. \end{cases}$$

This is the method of Eells and Sampson [4].

2. We shall have occasion to use a sharp form of the maximum principle. Let C denote a constant.

<u>Theorem.</u> Let f be a continuous function on $X \times [\alpha,\omega]$ with $f \leq 0$ on $X \times \alpha$ and on $\partial X \times [\alpha,\omega]$. Suppose that wherever $f > 0$ we could conclude that f is smooth and

$$\frac{\partial f}{\partial t} \leq \Delta f + Cf.$$

Then in fact we must always have $f \leq 0$.

Proof. Let $h = e^{-(C+1)t} f$. Then $h \leq 0$ on $X \times \alpha$ and $\partial X \times [\alpha,\omega]$. Moreover, wherever $h > 0$ we have $f > 0$ also, so h is smooth. Since

$$\frac{\partial h}{\partial t} = e^{-(C+1)t}f - (C+1)e^{-(C+1)t}f$$

and $\Delta h = e^{-(C+1)t}\Delta f$ we have

$$\frac{\partial h}{\partial t} \leq \Delta h - h.$$

Let (x,t) be a point where h assumes its maximum. If f is positive somewhere then $h(x,t) > 0$. We will show this leads to a contradiction. We must have $x \notin \partial X$ and $t > \alpha$. In local coordinates h is smooth at (x,t) and

$$\frac{\partial h}{\partial x^i}(x,t) = 0, \quad \frac{\partial^2 h}{\partial x^i \partial x^j}(x,t) \leq 0 \text{ as a matrix and } \frac{\partial h}{\partial t}(x,t) \geq 0.$$

Moreover,

$$\frac{\partial h}{\partial t} \leq g^{ij}\left\{\frac{\partial^2 h}{\partial x^i \partial x^j} - \Gamma_{ij}^k \frac{\partial h}{\partial x^k}\right\} - h.$$

But this implies $h(x,t) \leq 0$ which is a contradiction. This proves the theorem.

3. As an application we prove the following. Let X, Y and Z be compact Riemannian manifolds. We assume $Y \subseteq Z$, $\dim Y = \dim Z$ and Y is part of Z cut off by a smooth convex boundary ∂Y. Let $f: X \times [\alpha,\omega] \to Z$. We assume

(1) f is continuous and the first space derivatives $\frac{\partial f^\alpha}{\partial x^i}$ exist and are continuous on all of $X \times [\alpha,\omega]$.

(2) f is smooth in the interior and there it satisfies the heat equation $\frac{\partial f}{\partial t} = \Delta f$.

Theorem. If $f(X \times \alpha) \subseteq Y$ and $f(\partial X \times [\alpha,\omega]) \subseteq Y$ then $f(X \times [\alpha,\omega]) \subseteq Y$. Thus if a solution of the heat equation starts in Y and if the boundary remains in Y then the whole solution remains in Y, provided ∂Y is convex.

Proof. Let σ be a smooth function on Z which is negative inside Y and positive outside Y; moreover in a neighborhood N of ∂Y choose σ equal to the distance from ∂Y, negative inside and positive outside. Let $\rho = \sigma \circ f$. Then ρ is continuous on $X \times [\alpha,]$ and smooth in the interior. Also we have

$$\frac{\partial \rho}{\partial t} = \frac{\partial \sigma}{\partial y^\alpha} \frac{\partial f^\alpha}{\partial t}$$

$$\Delta\rho = \frac{\partial \sigma}{\partial y^\alpha} \Delta f^\alpha + g^{ij}\left[\frac{\partial^2 \sigma}{\partial y^\beta \partial y^\gamma} - \frac{\partial \sigma}{\partial y^\alpha} \Gamma^\alpha_{\beta\gamma}\right] \frac{\partial f^\beta}{\partial x^i} \frac{\partial f^\gamma}{\partial x^j} .$$

Therefore if f satisfies the heat equation $\frac{\partial f}{\partial t} = \Delta f$ we have

$$\frac{\partial \rho}{\partial t} = \Delta\rho - g^{ij}\left[\frac{\partial^2 \sigma}{\partial y^\beta \partial y^\gamma} - \frac{\partial \sigma}{\partial y^\alpha} \Gamma^\alpha_{\beta\gamma}\right] \frac{\partial f^\beta}{\partial x^i} \frac{\partial f^\gamma}{\partial x^j} .$$

In a neighborhood of a point on ∂Y we can choose local coordinates so that $Y = \{y_n \leq 0\}$ and $y_n = \sigma$. In such coordinates the condition that ∂Y is convex is that the matrix $\Gamma^n_{\beta\gamma} (1 \leq \beta, \gamma \leq n-1)$ be weakly negative-definite. Moreover we can choose our coordinate system so that the lines $y^1, \dots, y^{n-1}$ = constants, $y^n = t$ are geodesics perpendicular to ∂Y. In such coordinates $h_{nn} = 1$ so $\frac{\partial h_{nn}}{\partial y^\alpha} = 0$; and also $h_{n\beta} = 0$ along ∂Y for $\beta \neq n$. Therefore since

$$\Gamma^n_{n\alpha} = h^{n\beta} \cdot \frac{1}{2} \left(\frac{\partial h_{n\beta}}{\partial y^\alpha} + \frac{\partial h_{\alpha\beta}}{\partial y^n} - \frac{\partial h_{\alpha n}}{\partial y^\beta}\right)$$

we know that on ∂Y the only non-zero terms must have $\beta = n$, but then these are zero also. Thus $\Gamma^n_{n\alpha} = 0$ along ∂Y. Therefore in these coordinates the larger matrix $\Gamma^n_{\beta\gamma} (1 \leq \beta, \gamma \leq n)$ is also weakly negative-definite, i.e. $\Gamma^n_{\beta\gamma} \leq 0$. Write

$\sigma_{,\beta\gamma} = \frac{\partial^2\sigma}{\partial y^\beta \partial y^\gamma} - \frac{\partial\sigma}{\partial y^\alpha}\Gamma^\alpha_{\beta\gamma}$. Then the matrix $\sigma_{;\beta\gamma}(y)(1 \leq \beta, \gamma \leq n)$ is weakly negative-definite for $y \in \partial Y$. This provides an alternative characterization of the condition that ∂Y is convex.

Lemma. Let $\lambda(A)$ denote the largest eigenvalue of the symmetric matrix A. Then $\lambda(A)$ is Lipschitz continuous:

$$|\lambda(A)-\lambda(B)| \leq \| A-B\| .$$

Proof. Recall $\lambda(A) = \sup\{(Av,v): \|v\| \leq 1\}$. Now if $\|v\| \leq 1$

$$(Av,v) = (Bv,v) + ((A-B)v,v) \leq \lambda(B) + \|A-B\|.$$

Thus $\lambda(A) \leq \lambda(B) + \|A-B\|$. Likewise $\lambda(B) \leq \lambda(A) + \|A-B\|$. The result follows.

The same clearly applies to the smallest eigenvalue $\mu(A) = -\lambda(-A)$. Moreover $\sigma_{,\beta\gamma}(y)$ depends smoothly on y and hence

$$\|\sigma_{,\beta\gamma}(y) - \sigma_{,\beta\gamma}(y_0)\| \leq Cd(y,y_0).$$

Therefore $|\mu(\sigma_{,\beta\gamma}(y))-\mu(\sigma_{,\beta\gamma}(y_0))| \leq Cd(y,y_0)$. For each y there is a $y_0 \in \partial Y$ with $d(y,y_0) = \sigma(y)$ since $\sigma(y)$ is the distance to ∂Y. Thus

$$\mu(\sigma_{,\beta\gamma}(y)) \geq -C\sigma(y).$$

Moreover the derivatives $\frac{\partial f^\alpha}{\partial x^i}$ are continuous and hence remain bounded on $X \times [\alpha,\omega]$. Recalling that $\rho = \sigma \circ f$ we have shown that

$$\frac{\partial\rho}{\partial t} \leq \Delta\rho + C\rho$$

at all points x such that $f(x)$ lies in a certain neighborhood N of ∂Y. Suppose that the map $f\colon X \times [\alpha,\omega] \to Z$ does not stay entirely in Y. We can still choose β with $\alpha < \beta \leq \omega$ so that $f\colon X \times [\alpha,\beta] \to Z$ does not stay entirely in Y but does stay in $Y \cup N$. Then wherever $\rho > 0$ we have $\frac{\partial\rho}{\partial t} \leq \Delta\rho + C\rho$. Also $\rho \leq 0$

on $X \times \alpha$ and $\partial X \times [\alpha,\beta]$, since there f does map into Y. It follows from the maximum principle that $\rho \leq 0$ everywhere. If f does not stay in Y this gives a contradiction.

4. Next we prove uniqueness for solutions of the heat equation using the maximum principle. Let X and Y be compact Riemannian manifolds, possibly with boundaries. Let $f_1 : X \times [\alpha,\omega] \to Y$ and $f_2 : X \times [\alpha,\omega] \to Y$. We assume

1) f_1 and f_2 and their first space derivatives $\frac{\partial f_1^\alpha}{\partial x^i}$ and $\frac{\partial f_2^\alpha}{\partial x^i}$ exist and are continuous on all of $X \times [\alpha,\omega]$

2) f_1 and f_2 are smooth in the interior and there they satisfy the heat equation: $\frac{\partial f_1}{\partial t} = \Delta f_1$ and $\frac{\partial f_2}{\partial t} = \Delta f_2$.

<u>Theorem</u>. If $f_1 = f_2$ on $X \times \alpha$ and also on $\partial X \times [\alpha,\omega]$ then they agree on all of $X \times [\alpha,\omega]$.

<u>Proof</u>. Suppose not. The following arguments will hold on a tubular neighborhood N of the diagonal in $Y \times Y$. We can choose β with $\alpha < \beta \leq \omega$ such that the pair $(f_1(x,t), f_2(x,t)) \in N$ for all $(x,t) \in X \times [\alpha,\beta]$ but $f_1 \neq f_2$ on all of $X \times [\alpha,\beta]$. We will then derive a contradiction.

Let σ be the smooth function on N defined by $\sigma(y_1,y_2) = \frac{1}{2} d(y_1,y_2)^2$. Let $\rho(x,t) = \sigma(f_1(x,t), f_2(x,t))$. Then ρ is continuous on $X \times [\alpha,\beta]$ and smooth in the interior. Moreover we can compute

$$\frac{\partial \rho}{\partial t} = \Delta\rho - g^{ij}\left\{\frac{\partial^2\sigma}{\partial y_1^\beta \partial y_1^\gamma} - \frac{\partial\sigma}{\partial y_1^\alpha}\Gamma^\alpha_{\beta\gamma}(y_1)\right\}\frac{\partial f_1^\beta}{\partial x^i}\frac{\partial f_1^\gamma}{\partial x^j}$$
$$- 2g^{ij}\frac{\partial^2\sigma}{\partial y_1^\beta \partial y_2^\gamma}\frac{\partial f_1^\beta}{\partial x^i}\frac{\partial f_2^\gamma}{\partial x^j}$$
$$- g^{ij}\left\{\frac{\partial^2\sigma}{\partial y_2^\beta \partial y_2^\gamma} - \frac{\partial\sigma}{\partial y_2^\alpha}\Gamma^\alpha_{\beta\gamma}(y_2)\right\}\frac{\partial f_2^\beta}{\partial x^i}\frac{\partial f_2^\gamma}{\partial x^j}.$$

Let dim $Y = m$. We can regard the derivatives of σ as forming the coefficients of a symmetric $2m \times 2m$ matrix $M(y_1, y_2)$ depending smoothly on y_1 and y_2; this matrix is then applied to the 2m-vector $\left(\frac{\partial f_1^\alpha}{\partial x^i}, \frac{\partial f_2^\alpha}{\partial x^i}\right)$ $(1 \leq \alpha \leq m)$.

Let us write

$$M(y_1,y_2)(v_1,v_2) = \left\{\frac{\partial^2\sigma}{\partial y_1^\beta \partial y_1^\gamma} - \frac{\partial\sigma}{\partial y_1^\alpha}\Gamma^\alpha_{\beta\gamma}(y_1)\right\} v_1^\beta v_1^\gamma$$

$$- 2\left\{\frac{\partial^2\sigma}{\partial y_1^\beta \partial y_2^\gamma}\right\} v_1^\beta v_2^\gamma$$

$$+ \left\{\frac{\partial^2\sigma}{\partial y_2^\beta \partial y_2^\gamma} - \frac{\partial\sigma}{\partial y_2^\alpha}\Gamma^\alpha_{\beta\gamma}(y_2)\right\} v_2^\beta v_2^\gamma .$$

Choose coordinates at a point $x \in X$ so that at that point $g^{ij}(x) = \delta^{ij}$. Then at x,

$$\frac{\partial\rho}{\partial t} = \Delta\rho - \sum_i M(f_1,f_2)\left(\frac{\partial f_1}{\partial x^i}, \frac{\partial f_2}{\partial x^i}\right).$$

Choose geodesic coordinates centered at a point in Y midway between $f_1(x,t)$ and $f_2(x,t)$. Then $h_{\beta\gamma}(0) = \delta_{\beta\gamma}$ and $\Gamma^\alpha_{\beta\gamma}(0) = 0$. Since $\sigma(y_1,y_2) = \sigma(y_2,y_1)$ and $\sigma(y,y) = 0$ it follows that σ must have a power series expansion

$$\sigma(y_1,y_2) = \tfrac{1}{2}\delta_{\beta\gamma}(y_1^\beta - y_2^\beta)(y_1^\gamma - y_2^\gamma) + \lambda_{\beta\gamma\delta}(y_1^\beta - y_2^\beta)(y_1^\gamma - y_2^\gamma)(y_1^\delta + y_2^\delta) + O(y^4).$$

Then we can compute

$$\frac{\partial^2\sigma}{\partial y_1^\beta \partial y_1^\gamma}(w,-w) = \delta_{\beta\gamma} + \mu_{\beta\gamma\delta}w^\delta + O(w^2)$$

$$\frac{\partial^2\sigma}{\partial y_1^\beta \partial y_2^\gamma}(w,-w) = -\delta_{\beta\gamma} + O(w^2)$$

$$\frac{\partial^2\sigma}{\partial y_2^\beta \partial y_2^\gamma}(w,-w) = \delta_{\beta\gamma} - \mu_{\beta\gamma\delta}w^\delta + O(w^2)$$

for appropriate $\mu_{\beta\gamma\delta}$ symmetric in β and γ. Also $\frac{\partial\sigma}{\partial y_1^\alpha}(w,-w) = O(w)$ and $\Gamma^\alpha_{\beta\gamma}(w) = O(w)$ so $\frac{\partial\sigma}{\partial y_1^\alpha}(w,-w)\Gamma^\alpha_{\beta\gamma}(w) = O(w^2)$;

likewise $\frac{\partial\sigma}{\partial y_2^\alpha}(w,-w)\Gamma^\alpha_{\beta\gamma}(-w) = O(w^2)$. Therefore we have a power series expansion

$$M(w,-w)(v_1,v_2) = \delta_{\beta\gamma}(v_1^\beta - v_2^\beta)(v_1^\gamma - v_2^\gamma)$$
$$+ \mu_{\beta\gamma\delta}(v_1^\beta v_1^\gamma - v_2^\beta v_2^\gamma)w^\delta + O(w^2).$$

Now $\delta_{\beta\gamma}(v_1^\beta - v_2^\beta)(v_1^\gamma - v_2^\gamma) = |v_1 - v_2|^2$. We can factor $v_1^\beta v_1^\gamma - v_2^\beta v_2^\gamma = (v_1^\beta - v_2^\beta)v_1^\gamma + v_2^\beta(v_1^\gamma - v_2^\gamma)$ so that

$$|\mu_{\beta\gamma\delta}(v_1^\beta v_1^\gamma - v_2^\beta v_2^\gamma)w^\delta| \leq 2C|v_1 - v_2|(|v_1| + |v_2|)|w|.$$

Now

$$|v_1 - v_2|^2 - 2C|v_1 - v_2|(|v_1| + |v_2|)|w| + C^2(|v_1| + |v_2|)^2|w|^2 \geq 0.$$

Also the term $O(w^2)$ can be bounded by $C(|v_1|+|v_2|)^2|w|^2$. Therefore we obtain an estimate

$$M(w,-w)(v_1,v_2) \geq -C(|v_1|+|v_2|)^2|w|^2.$$

Now since our geodesic coordinates are centered midway between $f_1(x,t)$ and $f_2(x,t)$ we have $M(w,-w) = M(f_1,f_2)$ with $|w| = \frac{1}{2}d(f_1(x,t),f_2(x,t))$. Then

$$M(f_1,f_2)\left(\frac{\partial f_1}{\partial x^i}, \frac{\partial f_2}{\partial x^i}\right) \geq -C\left(\left|\frac{\partial f_1}{\partial x^i}\right| + \left|\frac{\partial f_2}{\partial x^i}\right|\right)^2|w|^2 .$$

But $\frac{\partial f_1}{\partial x^i}$ and $\frac{\partial f_2}{\partial x^i}$ are continuous and hence bounded on $X \times [\alpha,\beta]$. Also $|w|^2 = \frac{1}{2}\rho$. Therefore we have

$$\frac{\partial\rho}{\partial t} \leq \Delta\rho + C\rho.$$

From the maximum principle it follows that since $\rho = 0$ on $X \times \alpha$ and $\partial X \times [\alpha,\beta]$ we must have $\rho \leq 0$ on all of $X \times [\alpha,\beta]$. But $\rho = \frac{1}{2}d(f_1,f_2)^2$ so we must have $f_1 = f_2$ on all of $X \times [\alpha,\beta]$. This completes the proof.

5. Let X and Y be compact Riemannian manifolds with boundary, and assume ∂Y is convex. Form the double $\tilde{Y}$ of Y and extend the metric of Y smoothly to $\tilde{Y}$. Embed $\tilde{Y}$ in a suitable Euclidean space R^N. We do not take the ordinary Euclidean metric on R^N, however. Let T be a tubular neighborhood of $\tilde{Y}$ in R^N. Extend the metric on $\tilde{Y}$ smoothly to a metric on T. There is an involution $\iota : T \to T$ on the tubular neighborhood corresponding to multiplication by -1 in the fibers, having precisely Y for its fixed point set. We wish to choose the extension of the metric to T in such a way that $\iota : T \to T$ is an isometry; this can be accomplished by taking any extension and averaging it under the action of ι. Finally let B be a large ball (in the Euclidean metric) containing T, and extend the metric on T smoothly to all of R^N so as to equal the Euclidean metric outside of B. Then in the new metric ∂B is also convex.

If $f : X \to Y \subseteq B$ then we can form Δf as a map into Y or as a map into B. We denote these $\Delta_Y f$ and $\Delta_B f$ respectively. The next theorem shows the distinction to be unnecessary.

<u>Lemma</u>. If Y is a submanifold of Z and $f : X \to Y \subseteq Z$ then $\Delta_Y f = \pi_* \Delta_Z f$ where $\pi_* : TZ_y \to TY_y$ is the orthogonal projection of TZ_y onto the subspace TY_y.

<u>Proof</u>. Let $\dim Y = m$ and $\dim Z = k$. Choose coordinates $\{z^1, \ldots, z^m, z^{m+1}, \ldots, z^k\}$ around $y \in Y$ so that locally $Y = \{z^{m+1} = \ldots = z^k = 0\}$. Now

$$\Delta_Z f^\alpha = g^{ij} \left\{ \frac{\partial^2 f^\alpha}{\partial x^i \partial x^j} - \Gamma^k_{ij} \frac{\partial f^\alpha}{\partial x^k} + \Gamma^\alpha_{\beta\gamma} \frac{\partial f^\beta}{\partial x^i} \frac{\partial f^\gamma}{\partial x^j} \right\} .$$

Since $f : X \to Y$ we have $f^{m+1}, \ldots, f^n = 0$. Therefore $\Delta_Z f^\alpha = \Delta_Y f^\alpha$ for $\alpha = 1, \ldots, m$. Thus $\Delta_Y f = \pi_* \Delta_Z f$.

<u>Theorem</u>. If $f : X \to Y \subseteq B$ with the metric on B chosen as before then $\Delta_B f = \Delta_Y f$.

Proof. We know from the previous lemma that $\Delta_Y f = \pi_* \Delta_B f$. But $\iota : T \to T$ is an isometry so $\Delta_B f$ is invariant under ι. The fixed point set of ι is Y so the subspace of tangent vectors invariant under ι_* is precisely TY_y. Hence $\Delta_B f \in TY_y$ so $\pi_* \Delta_B f = \Delta_B f$. Thus $\Delta_B f = \Delta_Y f$.

Now let $f : X \times [\alpha,\omega] \to B$. We assume

(1) f is continuous and its first space derivatives $\frac{\partial f^\alpha}{\partial x^i}$ exist and are continuous on $X \times [\alpha,\omega]$.

(2) f is smooth in the interior of $X \times [\alpha,\omega]$ and there it satisfies the heat equation

$$\frac{\partial f}{\partial t} = \Delta_B f.$$

Theorem. If $f(X \times \alpha) \subseteq Y$ and $f(\partial X \times [\alpha,\omega]) \subseteq Y$ then $f(X \times [\alpha,\omega]) \subseteq Y$ and $f : X \times [\alpha,\omega] \to Y$ satisfies the heat equation

$$\frac{\partial f}{\partial t} = \Delta_Y f.$$

Proof. We proceed by contradiction. If the image of f does not always remain in Y, we can restrict ourselves to a smaller interval $X \times [\alpha,\beta]$ with $\alpha < \beta \leq \omega$ such that the image of f does not always remain in Y but does remain in the tubular neighborhood T of $\tilde{Y}$. Since $\iota : T \to T$ is an isometry, the map $\iota \circ f : X \times [\alpha,\beta] \to B$ is another solution of the heat equation. Since ι is the identity on Y this solution has the same initial and boundary values as f, so by our uniqueness theorem $\iota \circ f = f$. This shows that the image of f must remain in the fixed point set $\tilde{Y}$ of the involution ι. But now ∂Y is convex in $\tilde{Y}$, so the solution must remain in Y. Once we know that the solution is in Y we have $\Delta_B f = \Delta_Y f$. This completes the proof.

We use this result in the following way. We wish to show that the heat equation has a solution at least for a short period of time $[0,\varepsilon]$ and that the solution is smooth except at the corner $\partial X \times \alpha$.

These results are independent of the curvature assumption Riem $Y \leq 0$, which is used to show the solution exists for all time and converges as $t \to \infty$ to a harmonic map. Therefore in proving existence of a solution for a short time and regularity we may assume, without loss of generality, that the target is the ball B.

6. We shall need to make estimates in L^p_n for maps of one manifold into another. For the domain we take a compact foliated manifold with corners M; in our applictions $M = X \times [\alpha, \omega]$. For simplicity we assume the target manifold Y is embedded in a large ball B in a Euclidean space R^N. Then the map $f: M \to B \subseteq R^N$ can be represented as N ordinary functions $\{f^1, \ldots, f^N\}$ on M. Let $L^p_n(M; R^N) = \bigoplus_N L^p_n(M)$ be the N-fold direct sum of $L^p_n(M)$ with itself. When there is no possibility of confusion we shall sometimes write $L^p_n(M)$ or just L^p_n for $L^p_n(M; R^N)$. Also write $\mathcal{C}_0(M, B)$ for the space of continuous maps of M into B.

<u>Lemma</u>. Suppose $0 \leq r \leq s$ and $1 < p, q < \infty$. If $f \in \mathcal{C}_0(M; B)$ and $f \in L^q_s$ then $f \in L^p_r$ provided $pr < qs$. Moreover there exists a constant C independent of f such that

$$\| f \|_{L^p_r} \leq C \| f \|^{r/s}_{L^q_s} .$$

<u>Proof</u>. Let $v = \frac{pq(s-r)}{qs-pr} < \infty$. Since $f \in \mathcal{C}_0(M; B)$ surely $f \in L^v_0$ and $\| f \|_{L^v_0} \leq C$. Let $f = \{f^1, \ldots, f^N\}$. Then each $f^\alpha \in L^q_s \cap L^v_0$. By interpolation $f^\alpha \in L^p_r$ and

$$\| f^\alpha \|_{L^p_r} \leq C \| f^\alpha \|^{r/s}_{L^q_s} \| f^\alpha \|^{1-r/s}_{L^v_0} .$$

Thus $\| f \|_{L^p_r} \leq C \| f \|^{r/s}_{L^q_s}$.

Recall that a polynomial differential operator of type (n,k) for maps $f: M \to B$ is one which can be written locally as

$$Pf = \Sigma c_{\alpha_1 \cdots \alpha_\nu}(f)\, D^{\alpha_1} f \cdots D^{\alpha_\nu} f$$

with max $\| \alpha_i \| \leq k$ and $\Sigma \| \alpha_i \| \leq n$ where the coefficients $c_{\alpha_1 \cdots \alpha_\nu}(f)$ are smooth non-linear functions of f. Here $\| \alpha_i \|$ is taken with respect to the weights of the foliation of M. Thus if $M = X \times [\alpha, \omega]$ the operator $Pf = \frac{\partial f}{\partial t} - \Delta f$ is a typical example of a polynomial differential operator of type (2,2).

Theorem. Let P be a polynomial differential operator of type (n,k). Suppose $r \geq 0$ and $1 \leq p, q < \infty$. If $f \in \mathcal{C}_0(M,B)$ and $f \in L^q_s$ then $f \in L^p_r$ provided $r + k < s$ and $p(r+n) < qs$. Moreover there exists a constant C independent of f such that

$$\| Pf \|_{L^p_r} \leq C(1 + \| f \|_{L^q_s})^{q/p}.$$

Proof. Choose a finite altas of coordinate charts covering M and a partition of unity φ_i subordinate to the cover. Then $P = \Sigma \varphi_i P$ and we may regard the $\varphi_i P$ as defined in the Euclidean space which is the range of the coordinate chart. If the range is a half-space or a corner we can choose an extension operator E and extend not only f but also the coefficients of $\varphi_i P$. Then $\varphi_i P(f) = C\{E\varphi_i P(Ef)\}$ where C is a cutoff operator. Therefore it is sufficient to prove the estimate

$$\| Pf \|_{L^p_r} \leq C(1 + \| f \|_{L^q_s})^{q/p}$$

when P is a polynomial differential operator defined in a whole Euclidean space whose coefficients are smooth with compact support. We assume this from now on.

Choose ρ just slightly larger than r. Then we will have $\rho > r \geq 0$ and still have $\rho + k < s$ and $p(\rho+n) < qs$. Moreover we can take ρ to be not an integer. Then $L^p_r \subseteq \Lambda^p_\rho$ and $\|f\|_{L^p_r} \leq C\|f\|_{\Lambda^p_r}$ by section 6 of Part II. Hence it is sufficient to make the estimate in Λ^p_ρ instead. Now $\|g\|_{\Lambda^p_\rho}$ we recall is the least constant C such that $\|D^\gamma g\|_{L^p} \leq C$ when $\|\gamma\| < \rho$, and if $\rho-\sigma/\sigma_j < \|\gamma\| < \rho$ then $\|\Delta^v_j D^\gamma g\|_{L^p} \leq C|v|^{(\rho-\|\gamma\|)\sigma_j/\sigma}$. Take $g = Pf$. Since

$$Pf = \Sigma c_{\alpha_1\cdots\alpha_\nu}(f)\, D^{\alpha_1}f\cdots D^{\alpha_\nu}f$$

with $\max\|\alpha_i\| \leq k$ and $\Sigma\|\alpha_i\| \leq n$ we will have a formula

$$D^\gamma Pf = \Sigma c_{\beta_1\cdots\beta_\mu}(f)\, D^{\beta_1}f\cdots D^{\beta_\mu}f$$

with $\max\|\beta_i\| \leq k+\|\gamma\|$ and $\Sigma\|\beta_i\| \leq n+\|\gamma\|$ where the $c_{\beta_1\cdots\beta_\mu}(f)$ are some derivatives of the $c_{\alpha_1\cdots\alpha_\nu}(f)$ and hence are smooth with compact support. More precisely each $c(f) = c(x,f(x))$ where $c(x,y)$ is defined and smooth for all x in the Euclidean space of the coordinate chart and all $y \in B$ and vanishes for large x. Recalling that T^v_j is the translation operator and $\Delta^v_j = T^v_j - I$ we have a formula for the difference of a product

$$\Delta^v_j(fg) = \Delta^v_j f\cdot T^v_j g + f\cdot\Delta^v_j g$$

so we can write $\Delta^v_j D^\gamma Pf$ as a sum of terms of the form

$$\Delta^v_j\, c(f)\cdot T^v_j D^{\beta_1}f\cdots T^v_j\, D^{\beta_\mu}f$$

or

$$c(f)D^{\beta_1}f\cdots D^{\beta_{i-1}}f\cdot\Delta^v_j D^{\beta_i}f\cdot T^v_j D^{\beta_{i+1}}f\cdots T^v_j\, D^{\beta_\mu}f.$$

Choose $p_1,\ldots,p_\mu$ with p_i just a little less than $qs/\|\beta_i\|$. Since $qs/\|\beta_i\| > qs/(n+\rho) > p > 1$ we can still have $p_i > 1$. Next let $\theta = \rho-\|\gamma\|$ and choose $p'_0,p'_1,\ldots,p'_\mu$ with p'_0 just a little less than qs/θ and p'_i just a little less than $qs/(\|\beta_i\|+\theta)$. Again since

$$\Sigma \| \beta_i \| + \theta \leq n + \| \gamma \| + \rho - \| \gamma \| = n+\rho$$

we can still have p_0' and $p_i' > 1$. Then

$$\frac{1}{p_0'} + \frac{1}{p_1} + \dots + \frac{1}{p_\mu}$$

will be just a little more than $(\Sigma \| \beta_i \| + \theta)/qs$ and

$$(\Sigma \| \beta_i \| + \theta)/qs \leq (n+\rho)/qs < 1/p$$

so we can make

$$\frac{1}{p_0'} + \frac{1}{p_1} + \dots + \frac{1}{p_\mu} \leq \frac{1}{p} .$$

Likewise we can make

$$\frac{1}{p_0} + \frac{1}{p_1} + \dots + \frac{1}{p_{i-1}} + \frac{1}{p_i'} + \frac{1}{p_{i+1}} + \dots + \frac{1}{p_\mu} \leq \frac{1}{p}.$$

Moreover the terms we wish to estimate have fixed compact support so there is no problem if the sum is actually less. Therefore we have

$$\| c(f) D^{\beta_1} f \dots D^{\beta_\mu} f \|_{L^p} \leq C \| c(f) \|_{L^\infty} \| D^{\beta_1} f \|_{L^{p_1}} \dots \| D^{\beta_\mu} f \|_{L^{p_n}}$$

$$\| \Delta_j^v c(f) \cdot T_j^v D^{\beta_1} f \dots T_j^v D^{\beta_\mu} f \|_{L^p}$$
$$\leq C \| \Delta_j^v c(f) \|_{L^{p_0'}} \| D^{\beta_1} f \|_{L^{p_1}} \dots \| D^{\beta_\mu} f \|_{L^{p_n}}$$

$$\| c(f) D^{\beta_1} f \dots D^{\beta_{i-1}} f \cdot \Delta_j^v D^{\beta_i} f \cdot T_j^v D^{\beta_{i+1}} f \dots T_j^v D^{\beta_\mu} f \|_{L^p}$$
$$\leq C \| c(f) \|_{L^\infty} \| D^{\beta_1} f \|_{L^{p_1}} \dots \| D^{\beta_{i-1}} f \|_{L^{p_{i-1}}} \| \Delta_j^v D^{\beta_i} f \|_{L^{p_i'}}$$
$$\| D^{\beta_{i+1}} f \|_{L^{p_{i+1}}} \dots \| D^{\beta_\mu} f \|_{L^{p_\mu}} .$$

Now $p_i \| \beta_i \| < qs$ so by the previous lemma

$$\| D^{\beta_i} f \|_{L^{p_i}} \leq C \| f \|_{L^{p_i}_{\| \beta_i \|}} \leq C \| f \|_{L^q_s}^{\| \beta_i \| / s} .$$

Also by definition

$$\|\Delta_j^v D^{\beta_i} f\|_{L^{p_i'}} \leq C|v|^{\theta\sigma_j/\sigma} \|f\|_{\Lambda^{p_i'}_{\|\beta_i\|+\theta}}$$

Now $p_i'(\|\beta_i\|+\theta) < qs$ so again by the previous lemma

$$\|f\|_{\Lambda^{p_i'}_{\|\beta_i\|+\theta}} \leq C\|f\|_{L^{p_i'}_{\|\beta_i\|+\theta}} \leq C\|f\|_{L^q_s}^{(\|\beta_i\|+\theta)/s}$$

Since the coefficient $C(x,y)$ satisfies a Lipschitz condition

$$\|\Delta_j^v c(f)\|_{L^{p_0'}} \leq C|v|^{\theta} \|f\|_{\Lambda^{p_0'}_{\theta}}$$

and since $p_0'\theta < qs$, by the previous lemma

$$\|f\|_{\Lambda^{p_0'}_{\theta}} \leq C\|f\|_{L^{p_0'}_{\theta}} \leq C\|f\|_{L^q_s}^{\theta/s}.$$

Moreover for all of these inequalities we also have inclusions of the spaces involved. Therefore $f \in L^p_r$ and we have estimated each term by $C\|f\|_{L^q_s}^{(\Sigma\|\beta_i\|+\theta)/s}$. Since always $(\Sigma\|\beta_i\|+\theta)/s = (n+\rho)/s < q/p$ we have

$$\|f\|_{L^p_r} \leq C(1+\|f\|_{L^q_s})^{q/p}.$$

This completes the proof.

7. Consider $X \times [\alpha,\omega]$ as a foliated manifold with corners, with the foliation induced by the time function t projecting onto the second factor. Assign weight 2 to the x-variables and weight 1 to the t variable. Then the Laplacian Δ acting on ordinary functions by the formula

$$\Delta f = g^{ij}\left(\frac{\partial^2 f}{\partial x^i \partial x^j} - \Gamma^k_{ij}\frac{\partial f}{\partial x^k}\right) = g^{ij}f,_{ij}$$

is an elliptic operator on X, and the heat operator $A = \frac{\partial}{\partial t} - \Delta$ is a parabolic operator on $X \times [\alpha,\omega]$. Moreover either the Dirichlet condition $Bf = f|\partial X \times [\alpha,\omega]$ or the Neumann condition $Bf = \frac{\partial f}{\partial \nu}|\partial X \times [\alpha,\omega]$ is a parabolic complementing boundary condition. Here $\frac{\partial f}{\partial \nu}$ is the normal derivative. For locally $A(\xi,\theta) = \xi^2 + i\theta$ is zero only if $\theta = i\xi^2$, so there are no zeros of θ in the lower half-plane when ξ is real. This shows $A = \frac{\partial}{\partial t} - \Delta$ is parabolic. At the boundary $A(\xi,\eta,\theta) = \xi^2 + \eta^2 + i\theta$, so as a polynomial in η the roots occur at $\eta = \pm i\sqrt{\xi^2 + i\theta}$. Note that if ξ is real and $\operatorname{Im}\theta \leq 0$ then $\operatorname{Re}(\xi^2 + i\theta) \geq 0$ so $\operatorname{Re}\sqrt{\xi^2 + i\theta} \geq 0$. Therefore $\eta = + i\sqrt{\xi^2 + i\theta}$ is the root with positive imaginary part and so

$$A^+(\xi,\eta,\theta) = \eta - i\sqrt{\xi^2 + i\xi}\ .$$

For the Dirichlet boundary conditions $Bf = f|\partial X \times [\alpha,\omega]$ we have

$$B(\xi,\eta,\theta) = 1$$

and for the Neumann boundary conditions $Bf = \frac{\partial f}{\partial \nu}|\partial X \times [\alpha,\omega]$ we have

$$B(\xi,\eta,\theta) = \eta.$$

The first is clearly linearly independent mod A^+ since it is never zero and has lower degree. For the second,

$$\eta \equiv i\sqrt{\xi^2 + i\theta} \quad \mathrm{mod}\ \eta - i\sqrt{\xi^2 + i\theta}$$

and $\sqrt{\xi^2 + i\theta} = 0$ only when $\theta = i\xi^2$, so there are no solutions for real ξ and $\operatorname{Im}\theta \leq 0$ except $\xi = \theta = 0$. Thus the second is also linearly independent mod A^+, so either is a complementing boundary condition.

We observe also that, at least away from the corner $\partial X \times \alpha$, we can regard the initial value problem $\frac{\partial f}{\partial t} = \Delta f$ on $X \times [\alpha,\omega]$ and $f = k$ on $X \times \alpha$ as a semi-elliptic equation with complementing boundary condition. For the operator $A = \frac{\partial}{\partial t} - \Delta$ has symbol $\xi^2 + i\theta$, which as a polynomial in θ has just one root, which has positive imaginary part, namely $\theta = i\xi^2$. Thus $A^+(\xi,\theta) = \theta - i\xi^2$. The boundary operator $Bf = f|X \times \alpha$ has symbol 1 which is surely linearly independent modulo $\theta - i\xi^2$ as a polynomial in θ. At the terminal boundary the operator $\frac{\partial f}{\partial t} - \Delta f$ is also semi-elliptic, but in local coordinates its symbol is $\xi^2 - i\theta$ since the time axis must be put the other way to agree with our conventions. Then there are zero roots with positive imaginary part, so (although it may sound paradoxical the empty set of boundary conditions are complementing.

8. We can therefore apply the results of Part III to the linear heat equation $\frac{\partial f}{\partial t} - \Delta f = g$ with Dirichlet or Neumann boundary conditions. For reference we state this explicitly.

Regard $X \times [\alpha,\omega]$ as a foliated manifold with corners. Then $X \times \alpha$ is a proper part of the boundary. We write $L_n^p(X \times [\alpha,\omega]/\alpha)$ for $L_n^p(X \times [\alpha,\omega]/X \times \alpha)$ defined as in Part II. Likewise we write $\partial L_n^p(\partial X \times [\alpha,\omega]/\alpha)$ for $\partial L_n^p(\partial X \times [\alpha,\omega]/\partial X \times \alpha)$, the space of boundary values of functions in $L_n^p(X \times [\alpha,\omega]/\alpha)$ on the proper part of the boundary $\partial X \times [\alpha,\omega]$.

<u>Theorem</u>. Let $1 < p < \infty$ and $n > 1/p$. For every $g \in L_{n-2}^p(X \times [\alpha,\omega]/\alpha)$ and $h \in \partial L_n^p(\partial X \times [\alpha,\omega]/\alpha)$ there exists a unique $f \in L_n^p(X \times [\alpha,\omega]/\alpha)$ with $\frac{\partial f}{\partial t} - \Delta f = g$ and $f|\partial X \times [\alpha,\omega] = h$.

<u>Theorem</u>. Let $1 < p < \infty$ and $n > 1 + 1/p$. For every $g \in L_{n-2}^p(X \times [\alpha,\omega]/\alpha)$ and $h \in \partial L_{n-1}^p(\partial X \times [\alpha,\omega]/\alpha)$ there exists a unique $f \in L_n^p(X \times [\alpha,\omega]/\alpha)$ with $\frac{\partial f}{\partial t} - \Delta f = g$ and $\frac{\partial f}{\partial \nu}|\partial X \times [\alpha,\omega] = h$.

Notice that the initial condition is stated implicitly by requiring all the functions in effect to vanish to $t \leq \alpha$. We can rephrase this as a solution of an initial-boundary value problem, but we cannot expect to obtain regularity at the corner.

Theorem. Let g be smooth on $X \times [\alpha, \omega]$, let h be smooth on $\partial X \times [\alpha, \omega]$ and let k be smooth on $X \times \alpha$. We require that $h = k$ on the corner $\partial X \times \alpha$. Then there exists a continuous function f on $X \times [\alpha, \omega]$ with continuous first space derivatives $\partial f/\partial x^i$ and smooth except at the corner $\partial X \times \alpha$ with $\frac{\partial f}{\partial t} - \Delta f = g$, $f|\partial X \times [\alpha, \omega] = h$ and $h|X \times \alpha = k$.

Proof. Since $h = k$ on $\partial X \times \alpha$ we can choose a smooth function f_0 equal to h on $\partial X \times [\alpha, \omega]$ and equal to k on $X \times \alpha$. Choose $p > \dim X + 2$ and regard $g - \frac{\partial f_0}{\partial t} + \Delta f_0$ as an element of $L_0^p(X \times [\alpha, \omega]/\alpha)$. Then we can find an $f_1 \in L_2^p(X \times [\alpha, \omega]/\alpha)$ with $\frac{\partial f_1}{\partial t} - \Delta f_1 = g - \frac{\partial f_0}{\partial t} + \Delta f_0$ on $X \times [\alpha, \omega]$ and $f_1 = 0$ on $\partial X \times [\alpha, \omega]$. Since $f_1 \in L_2^p(X \times [\alpha, \omega]/\alpha)$ we automatically have f_1 continuous and $\partial f_1/\partial x^i$ continuous and $f_1 = 0$ on $X \times \alpha$. Moreover, our regularity results for semi-elliptic equations show f_1 is smooth except at the corner $\partial X \times \alpha$. Then $f = f_0 + f_1$ is the desired solution. In the same way we prove the following:

Theorem. Let g be smooth on $X \times [\alpha, \omega]$ let h be smooth on $\partial X \times [\alpha, \omega]$ and let k be smooth on $X \times \alpha$. We require that $\frac{\partial k}{\partial \nu} = h$ on the corner $\partial X \times \alpha$. Then there exists a continuous function f on $X \times [\alpha, \omega]$ whose space derivatives $\frac{\partial f}{\partial x^i}$ exist and are continuous everywhere and such that f is smooth except at the corner $\partial X \times \alpha$, with $\frac{\partial f}{\partial t} - \Delta f = g$ on $X \times [\alpha, \omega]$, $\frac{\partial f}{\partial \nu} = h$ on $\partial X \times [\alpha, \omega]$ and $f = k$ on $X \times \alpha$.

The argument is unchanged if we add on terms involving lower order derivatives. For example, we use the following in Part V. Let a be a smooth vector field on X and write $a\nabla f = a^i\, \partial f/\partial x^i$.

Let b be a smooth function on X and c a smooth function on ∂X.

Theorem. Let g be smooth on $X \times [\alpha,\omega]$, let h be smooth on $\partial X \times [\alpha,\omega]$, and let k be smooth on $X \times \alpha$. We require that $\frac{\partial k}{\partial \nu} + ck = h$ on the corner $\partial X \times \alpha$. Then there exists a continuous function f on $X \times [\alpha,\omega]$ with continuous first space derivatives $\nabla f = \partial f/\partial x^i$, and with f smooth except at the corner $\partial X \times \alpha$, satisfying

$$\frac{\partial f}{\partial t} - (\Delta f + a\nabla f + bf) = g \quad \text{on} \quad X \times [\alpha,\omega]$$

$$\frac{\partial f}{\partial \nu} - cf = h \quad \text{on} \quad \partial X \times [\alpha,\omega]$$

$$f = k \quad \text{on} \quad X \times \alpha .$$

9. We adopt the following notation. Write ∇f for the matrix of first space derivatives $\frac{\partial f^\alpha}{\partial x^i}$. The Laplacian

$$\Delta f^\alpha = g^{ij}\left\{\frac{\partial^2 f^\alpha}{\partial x^i \partial x^j} - \frac{\partial f^\alpha}{\partial x^k}\Gamma^k_{ij} + \Gamma^\alpha_{\beta\gamma}(f)\frac{\partial f^\beta}{\partial x^i}\frac{\partial f^\gamma}{\partial x^j}\right\}.$$

Regarding f^α as a real-valued function, its ordinary Laplacian is

$$\Delta f^\alpha = g^{ij}\left\{\frac{\partial^2 f^\alpha}{\partial x^i \partial x^j} - \frac{\partial f^\alpha}{\partial x^k}\Gamma^k_{ij}\right\}.$$

Write symbolically $\Delta f = \{\Delta f^\alpha\}$ and

$$\Gamma(f)\nabla f^2 = g^{ij}\,\Gamma^\alpha_{\beta\gamma}(f)\frac{\partial f^\beta}{\partial x^i}\frac{\partial f^\gamma}{\partial x^j}.$$

Then we have

$$\Delta f = \Delta f + \Gamma(f)\nabla f^2 .$$

In case the target is the ball B in a Euclidean space we have a global system of coordinates on B and the decomposition $\blacktriangle f = \Delta f + \Gamma(f)\nabla f^2$ makes sense globally; however only $\blacktriangle f$ transforms properly under changes of coordinates on the target. Observe that $\Gamma(f)\nabla f^2$ is a polynomial differential operator of type $(2,1)$ on X or on $X \times [\alpha,\omega]$.

Suppose $f \in L^p_2(X \times [\alpha,\omega])$ with $p > \dim X+2$. Then by the Sobolev Embedding Theorem f and $\nabla f = \{\partial f^\alpha/\partial x^i\}$ are continuous so it makes sense to say $f:X \times [\alpha,\omega] \to B$. Also $\Gamma(f)\nabla f^2$ is defined and continuous while $\frac{\partial f}{\partial t}$ and Δf are defined and in $L^p(X \times [\alpha,\omega])$. Therefore it makes sense to say that $f \in L^p_2$ with $p > \dim X+2$ satisfies the heat equation $\frac{\partial f}{\partial t} = \blacktriangle f$. We now prove a regularity theorem.

<u>Theorem</u>. Let $f:X \times [\alpha,\omega] \to B$. Suppose $f \in L^p_2(X \times [\alpha,\omega])$ and satisfies the heat equation $\frac{\partial f}{\partial t} = \blacktriangle f$. Suppose also that $f|X \times \alpha$ and $f|\partial X \times [\alpha,\omega]$ are smooth. Then f is smooth on $X \times [\alpha,\omega]$ except possibly at the corner $\partial X \times \alpha$.

<u>Proof</u>. First we show f is smooth for $t > \alpha$. Let $h = f|\partial X \times [\alpha,\omega]$. Then $\frac{\partial f^\alpha}{\partial t} - \Delta f^\alpha$ is a parabolic operator and $f^\alpha|\partial X \times [\alpha,\omega]$ is a parabolic complementing boundary condition. Moreover $f^\alpha|\partial X \times [\alpha,\omega]$ is always smooth. Therefore the regularity result for parabolic equations in Part III tells us that if $\alpha \leq \beta < \gamma < \omega$ and $\Gamma(f)\nabla f^2 \in L^u_n(X \times [\beta,\omega])$ then $f \in L^u_{n+2}(X \times [\gamma,\omega])$ for all n and u with $1 < u < \infty$. As soon as $f \in L^p_2(X \times [\alpha,\omega])$ for some $p > \dim X+2$ we have $\Gamma(f)\nabla f^2$ continuous so $f \in L^u_2(X \times [\beta,\omega])$ for every $\beta > \alpha$ and every $u < \infty$.

Moreover $\Gamma(f)\nabla f^2$ is a polynomial differential operator of type $(2,1)$ so by Section 6 if $f \in L^u_s$ then $\Gamma(f)\nabla f^2 \in L^v_r$ provided $r+1 < s$ and $v(r+2) < us$. If u can be made arbitrarily large then the last restriction is superfluous and v can be made arbitrarily large as well. For simplicity let us say $f \in L^*_s$ if

$f \in L_s^u(X \times [\beta,\omega])$ for every $\beta > \alpha$ and $u < \infty$. Then we have shown the following facts:

1) $f \in L_2^*$.

2) $\Gamma(f)\nabla f^2 \in L_n^* \Longrightarrow f \in L_{n+2}^*$.

3) $f \in L_s^* \Longrightarrow \Gamma(f)\nabla f^2 \in L_r^*$ if $r + 1 < s$.

It follows by induction in increments less than 1 that $f \in L_n^*$ for all n. Hence f is smooth for $t > \alpha$.

Along the initial boundary $X \times \alpha$ but away from the corner we can apply a similar argument using $\frac{\partial f}{\partial t} - \Delta f$ as a semi-elliptic operator and $f|X \times \alpha$ as a complementing boundary condition. This shows that f is smooth except possibly at the corner $X \times \alpha$. In fact it will not be smooth there in general unless an infinite number of compatibility conditions between derivatives of $f|X \times \alpha$ and $f|\partial X \times [\alpha,\omega]$ are satisfied at the corner.

10. For the next section we shall need to consider a linear parabolic system. All of the results of Part III for semi-elliptic and parabolic equations extend to systems in the same way as is done for elliptic equations in Agmon, Douglis and Nirenberg II [1]. However, the simple case we need can more easily be derived from the theory for a single function. The system we wish to consider is the weakly-coupled heat equation

$$\frac{\partial f}{\partial t} = \Delta f + a\nabla f + bf$$

where $f = \{f^1,\ldots,f^N\}$, $(a\nabla f)^\alpha = a_\beta^{\alpha i}\frac{\partial f^\beta}{\partial x^i}$ and $(bf)^\alpha = b_\beta^\alpha f^\beta$. We take the coefficients $a_\beta^{\alpha i}$ and b_β^α to be smooth, and choose $p > \dim X + 2$.

<u>Theorem.</u> For every $g \in L_0^p(X\times[\alpha,\omega]/\alpha)$ and $h \in \partial L_2^p(\partial X\times[\alpha,\omega]/\alpha)$ there exists a unique $f \in L_2^p(X\times[\alpha,\omega]/\alpha)$ with

$$\frac{\partial f}{\partial t} - \Delta f - a\,\nabla f - bf = g$$

on $X \times [\alpha,\omega]$ and $f|\partial X \times [\alpha,\omega] = h$.

Remark. Of course $L_0^p(X\times[\alpha,\omega]/\alpha) = L^p(X\times[\alpha,\omega])$.

Proof. Let $Hf = \frac{\partial f}{\partial t} - \Delta f$ and $Kf = a\nabla f + bf$. By the theory for a single equation repeated N times, the map $f \to (Hf, f|\partial X\times[\alpha,\omega])$ defines an isomorphism of $L_2^p(X\times[\alpha,\omega]/\alpha)$ onto $L_0^p(X\times[\alpha,\omega]/\alpha) \oplus \partial L_2^p(\partial X\times[\alpha,\omega]/\alpha)$. Moreover, since K has weight 1 the map $K\colon L_2^p \to L_0^p$ is compact, since it factors through the compact inclusion of L_1^p into L_0^p. By the theory of Fredholm mappings the map $L_2^p \to L_0^p \oplus \partial L_2^p$ given by $f \to (Hf-Kf, f|\partial X\times[\alpha,\omega])$ has finite dimensional kernel and cokernel. Moreover its index is zero. Therefore to show that it is an isomorphism it suffices to show that its kernel is zero.

Let $f \in L_2^p(X\times[\alpha,\omega]/\alpha)$ and suppose $\frac{\partial f}{\partial t} = \Delta f + a\nabla f + b$ and $f|\partial X\times[\alpha,\omega] = 0$. Since $p > \dim X + 2$ we know that f and $\nabla f = \{\partial f^\alpha/\partial x^i\}$ are continuous, and since $f \in L_2^p(X\times[\alpha,\omega]/\alpha)$ we know $f|X \times \alpha = 0$. Using the regularity properties for the single equation

$$\frac{\partial f^\alpha}{\partial t} - \Delta f^\alpha = a_\beta^{\alpha i}\frac{\partial f^\beta}{\partial x^i} + b_\beta^\alpha f^\beta$$

for each α an easy induction shows that the f^α are smooth for $t > \alpha$. Let $\chi = \frac{1}{2}|f|^2 = \frac{1}{2}\sum_{\alpha=1}^{n}(f^\alpha)^2$. Then by an easy computation

$$\frac{\partial \chi}{\partial t} = \Delta\chi - |\nabla f|^2 + a\cdot f\cdot\nabla f + bf^2$$

where $af\nabla f = a_\beta^{\alpha i} f^\alpha \frac{\partial f^\beta}{\partial x^i}$ and $bf^2 = b_\beta^\alpha f^\alpha f^\beta$. Then for an appropriate constant C we have

$$-|\nabla f|^2 + af\nabla f + bf^2 \leq \frac{1}{2}Cf^2 = C\chi$$

and

$$\frac{\partial \chi}{\partial t} \leq \Delta\chi + C\chi.$$

Since $\chi = 0$ on $X \times \alpha$ and $\partial X \times [\alpha,\omega]$ the maximum principle shows that $\chi = 0$. Thus $f = 0$. This proves the theorem.

11. We now prove existence of solutions of the non-linear heat equation $\frac{\partial f}{\partial t} = \text{A}f$ for short periods of time. Let X be a compact Riemannian manifold and choose a Riemannian metric on the Euclidean space R^N equal to the usual metric outside a large ball B. Choose $p > \dim X + 2$.

Theorem. Let $f_\alpha\colon X \to B$ and $h\colon \partial X \times [\alpha,\omega] \to B$ be smooth maps with $f_\alpha = h$ on the corner $\partial X \times \alpha$. There exists an $\epsilon > 0$ (depending on f_α and h) and a map $f\colon X \times [\alpha,\alpha+\epsilon] \to B$ with $f \in L_2^p(X\times[\alpha,\alpha+\epsilon])$ solving the equation

$$\frac{\partial f}{\partial t} = \text{A}f \qquad \text{on } X \times [\alpha,\alpha+\epsilon]$$

$$f = f_\alpha \qquad \text{on } X \times \alpha$$

$$f = h \qquad \text{on } \partial X \times [\alpha,\alpha+\epsilon].$$

Moreover, f is unique, and smooth except on the corner $\partial X \times \alpha$.

Proof. We use the classical inverse function theorem on Banach spaces (see Lang [13]). We will find f as a sum $f_\flat + f_\#$ where $f_\flat$ satisfies the boundary conditions and $f_\#$ is a correction. Choose $f_\flat : X \times [\alpha,\omega] \to R^N$ to be a smooth map with $f_\flat = f_\alpha$ on $X \times \alpha$ and $f_\flat = h$ on $\partial X \times [\alpha,\omega]$. This is possible since $f_\alpha = h$ on the corner $\partial X \times \alpha$. The derivative of a non-linear operator Pf in the direction k is its first variation

$$DP(f)k = \lim_{\theta\to 0}[P(f+\theta k)-P(f)]/\theta\ .$$

Consider the operator

$$\text{A}f = \Delta f + \Gamma(f)\nabla f^2.$$

Its derivative is given by the formula

$$D\text{A}(f)k = \Delta k + D\Gamma(f)k\cdot\nabla f^2 + 2\Gamma(f)\nabla f\cdot\nabla k.$$

To be precise, in local coordinates

$$\Delta f^{\alpha} = g^{ij}\{\frac{\partial^2 f^{\alpha}}{\partial x^i \partial x^j} - \frac{\partial f^{\alpha}}{\partial x^{\ell}} \Gamma^{\ell}_{ij} + \Gamma^{\alpha}_{\beta\gamma}(f) \frac{\partial f^{\beta}}{\partial x^i} \frac{\partial f^{\gamma}}{\partial x^j}\}$$

$$D\Delta(f)k^{\alpha} = g^{ij}\{\frac{\partial^2 k^{\alpha}}{\partial x^i \partial x^j} - \frac{\partial k^{\alpha}}{\partial x^{\ell}} \Gamma^{\ell}_{ij}\} + g^{ij} \frac{\partial \Gamma^{\alpha}_{\beta\gamma}}{\partial y^{\delta}} k^{\delta} \frac{\partial f^{\beta}}{\partial x^i} \frac{\partial f^{\gamma}}{\partial x^j}$$

$$+ 2g_{ij}\Gamma^{\alpha}_{\beta\gamma}(f)\frac{\partial f^{\beta}}{\partial x^i} \frac{\partial k^{\gamma}}{\partial x^j} .$$

The important observation here is that if f is smooth the operator $D\Delta(f)k$ has the form

$$D\Delta(f)k = \Delta k + a\nabla k + bk$$

where a and b are smooth matrices of functions, as in Section 10. Let $H(f) = \frac{\partial f}{\partial t} - \Delta f$. Then $DH(f)k = \frac{\partial k}{\partial t} - \Delta k - a\nabla k - bk$. Fix $f_\flat$ and consider $f_\#$ as the variable function. Let $L^p_2(X\times[\alpha,\omega]/\alpha)_\#$ denote the closed linear subspace of those $f_\# \in L^p_2(X\times[\alpha,\omega]/\alpha)$ with $f_\# | \partial X \times [\alpha,\omega] = 0$. We will also have $f_\# | X \times \alpha = 0$ by the definition of $L^p_2(X\times[\alpha,\omega]/\alpha)$. Then $f_\# \to H(f_\flat + f_\#)$ defines a continuously differentiable map of $L^p_2(X\times[\alpha,\omega]/\alpha)_\# \to L^p(X\times[\alpha,\omega])$. Its derivative at $f_\# = 0$ is

$DH(f_\flat)\colon L^p_2(X\times[\alpha,\omega]/\alpha)_\# \to L^p(X\times[\alpha,\omega])$

$DH(f_\flat)k = \frac{\partial k}{\partial t} - \Delta k - a\cdot\nabla k - bk$

which by the theorem in Section 10 is an isomorphism. Therefore by the inverse function theorem the set of all $H(f_\flat + f_\#)$ for $f_\#$ in a neighborhood of 0 covers a neighborhood of $H(f_\flat)$ in $L^p(X\times[\alpha,\omega])$. If we choose $\varepsilon > 0$ small enough, the function equal to 0 for $\alpha \leq t \leq \alpha+\varepsilon$ and equal to $H(f_\flat)$ for $\alpha+\varepsilon < t \leq \omega$ will be in this neighborhood. Therefore we can choose $f_\# \in L^p_2(X\times[\alpha,\omega]/\alpha)_\#$ with $H(f_\flat + f_\#) = 0$ on $X \times [\alpha,\alpha+\varepsilon]$. Let $f = f_\flat + f_\#$. Then $f = f_\alpha$ on $X \times \alpha$ and $f = h$ on $\partial X \times [\alpha,\omega]$ and $f \in L^p_2(X\times[\alpha,\omega])$ satisfies $\frac{\partial f}{\partial t} = \Delta f$ at least on $X \times [\alpha,\alpha+\varepsilon]$.

Since f is continuous the solution lies in some ball B'. By regularity the solution is smooth except at the corner $\partial X \times \alpha$, and since ∂B is convex the solution f remains inside B. Also by a previous theorem the solution is unique. This completes the proof.

Let Y be a compact manifold and $f: X \times [\alpha,\omega] \to Y$. If $p > \dim X+2$ then the condition that f be of class $L_2^p(X\times[\alpha,\omega];Y)$ is invariantly defined. That is, we require that the representatives f^β be of class $L_2^p(X\times[\alpha,\omega])$ for an atlas of coordinate charts on Y. If f^β are the representatives in coordinates y^β and $\tilde{f}^\beta$ those in coordinates $\tilde{y}^\beta$ then we have

$$\frac{\partial \tilde{f}^\beta}{\partial x^i} = \frac{\partial \tilde{y}^\beta}{\partial y^\gamma}\frac{\partial f^\gamma}{\partial x^i} \qquad \frac{\partial \tilde{f}^\beta}{\partial t} = \frac{\partial \tilde{y}^\beta}{\partial y^\gamma}\frac{\partial f^\gamma}{\partial t}$$

$$\frac{\partial^2 \tilde{f}^\beta}{\partial x^i \partial x^j} = \frac{\partial \tilde{y}^\beta}{\partial y^\gamma}\frac{\partial^2 f^\gamma}{\partial x^i \partial x^j} + \frac{\partial^2 \tilde{y}^\beta}{\partial x^\gamma \partial y^\delta}\frac{\partial f^\gamma}{\partial x^i}\frac{\partial f^\delta}{\partial x^j}.$$

The $f^\gamma \in L_2^p(X\times[\alpha,\omega])$ if and only if f^γ, $\frac{\partial f^\gamma}{\partial x^i}$, $\frac{\partial^2 f^\gamma}{\partial x^i \partial x^j}$ and $\frac{\partial f^\gamma}{\partial t}$ belong to $L^p(X\times[\alpha,\omega])$. But if $p > \dim X + 2$ then f^γ and $\frac{\partial f^\gamma}{\partial x^i}$ are continuous, and we can conclude that the $\tilde{f}^\beta \in L_2^p(X\times[\alpha,\omega])$ as well. For lower values of p there are problems.

Corollary. Let X and Y be compact Riemannian manifolds with boundary and suppose ∂Y is convex. Choose $p > \dim X + 2$. If $f_0: X \times \alpha \to Y$ and $h: \partial X \times [\alpha,\omega] \to Y$ are smooth and $f_0 = h$ on the corner $\partial X \times \alpha$ then there exists an $\epsilon > 0$ and a map $f: X \times [\alpha,\alpha+\epsilon] \to Y$ of class $L_2^p(X\times[\alpha,\alpha+\epsilon];Y)$ satisfying the heat equation

$$\frac{\partial f}{\partial t} = \Delta f \quad \text{on } X \times [\alpha,\alpha+\epsilon]$$

$$f = f_0 \quad \text{on } X \times \alpha$$

$$f = h \quad \text{on } \partial X \times [\alpha,\alpha+\epsilon].$$

Moreover f is unique, and smooth except on the corner $\partial X \times \alpha$.

Proof. Embed Y in B as in Section 5 and apply the previous theorem.

Part V: Growth Estimates and Convergence

1. In this part of the paper we prove growth estimates on the solution of the heat equation which show that the solution exists for all time and converges as $t \to \infty$ to a harmonic map. These estimates depend strongly on the curvature assumption Riem $Y \leq 0$, which we have not used until now. Let X and Y be compact Riemannian manifolds with ∂Y convex. Let $h:\partial X \to Y$ be a given smooth map, and suppose we seek a harmonic map with boundary value h in a given relative homotopy class. Let $f_0:X \to Y$ be a smooth map in that relative homotopy class with boundary value $f_0|\partial X = h$. Let $f:X \times [0,\omega) \to Y$ be a solution locally of class L_2^p for some $p > \dim X+2$ of the heat equation

$$\frac{\partial f}{\partial t} = \Delta f \quad \text{on} \quad X \times [0,\omega)$$

$$f = f_0 \quad \text{on} \quad X \times 0$$

$$f = h \quad \text{on } \partial X \times [0,\omega).$$

We know such a solution exists on $X \times [0,\varepsilon]$ for some $\varepsilon > 0$, and is smooth except at the corner $\partial X \times 0$. Moreover two such solutions must agree as long as both are defined. Since a union of intervals $[0, \omega_i)$ is an interval $[0,\omega)$ we may take $\omega \leq \infty$ to be as large as possible. At least $\omega > 0$. Note that here the boundary condition $f = h$ on $\partial X \times [0,\omega)$ is independent of time since $h(x)$ depends only on $x \in \partial X$ and not on t.

Choose $\delta > 0$ with $\delta = 1$ if $\omega = \infty$ and small compared to ω if $\omega < \infty$; $\delta < \omega/4$ will do. Let $\|f\|_\tau$ and $\|f\|_{[\tau,\tau+\delta]}$ measure various norms of f on $X \times \tau$ or $X \times [\tau,\tau+\delta]$ respectively. In the following arguments we will show that various $\|f\|_\tau$ or $\|f\|_{[\tau,\tau+\delta]}$ are uniformly bounded independently of τ, for $0 \leq \tau < \omega$ in the first case and $0 \leq \tau < \omega-\delta$ in the second. These will be used in two ways.

(1) If $\omega < \infty$, we conclude that the $f_t(x) = f(x,t)$ converge to a map $f_\omega(x)$ as $t \to \omega$, extending the solution f to the closed set $X \times [0,\omega]$. We then use the existence theorem to extend the solution to $X \times [\omega,\omega+\epsilon]$ for some $\epsilon > 0$. This will show ω is not maximal unless $\omega = \infty$.

(2) If $\omega = \infty$, we conclude that (at least a subsequence of) the $f_t(x)$ converge to a harmonic map $f_\infty(x)$ as $t \to \infty$ with boundary value h and in the given relative homotopy class.

2. The curvature of Y enters in the computation of parabolic inequalities for the potential energy $\chi = \frac{1}{2}|\nabla f|^2$ and the kinetic energy $\varkappa = \frac{1}{2}\left|\frac{\partial f}{\partial t}\right|^2$. The interior inequalities are proved in Eells and Sampson [4]. We also compute inequalities for the normal derivatives $\frac{\partial \chi}{\partial \nu}$ and $\frac{\partial \varkappa}{\partial \nu}$. The philosophy is that in physics $\frac{\partial}{\partial t} - \Delta$ represents the rate of generation of heat in the interior while $\frac{\partial}{\partial \nu}$ represents the heat flow across the boundary. Mathematically the idea is to derive a polynomial partial differential equation or inequality whose highest order terms are linear.

We present the computation both in the classical notation of coordinate systems and the modern coordinate-free notation. In the classical notation we use $i,j,k \ldots$ for indices on X and $\alpha,\beta,\gamma,\ldots$ for indices on Y. We let g_{ij} and $h^{\alpha\beta}$ be the Riemannian metrics on X and Y, Γ^i_{jk} and $\Gamma^\alpha_{\beta\gamma}$ the Christoffel symbols, R^{ij} the Ricci curvature on X and $R_{\alpha\beta\gamma\delta}$ the Riemannian curvature on Y. In the modern notation we use $v,w,\ldots$ for tangent vectors on X, ∇ for covariant derivative along X of a section of a vector bundle, ∇_v for the covariant derivative in the direction v, and $\frac{\partial}{\partial t}$ for derivative in the time direction. Because $X \times [0,\omega)$ has a product structure ∇ commutes with $\frac{\partial}{\partial t}$. We use $\langle\ ,\ \rangle$ for the inner product in a Riemannian bundle, in particular f^*TY. As usual

$|g|^2 = \langle g,g \rangle$. We shall also need to have contractions of tensors in different arguments, and we denote this by the convention that when a vector field $\dot{v}$ is repeated in the notation we have contracted on this argument. Thus

$$\Delta f = \operatorname{Tr} \nabla\nabla f = \nabla_v \nabla_v f.$$

If $f: X \times [0,\infty) \to Y$ we let $\nabla f_{(x,t)}: TX \to TY$ denote its total X-derivative as a linear map. We can regard ∇f as a section of the bundle $L(TX, f^*TY)$. In local coordinates $\nabla f = \{\frac{\partial f^\alpha}{\partial x^i}\}$. It has a covariant derivative $\nabla\nabla f$ which is a section of the bundle $L_S^2(TX, f^*TY)$ of symmetric bilinear maps of TX into f^*TY. The covariant derivative is taken with respect to the natural connection on $L(TX, f^*TY)$, which is minus the connection on TX plus that on f^*TY. In local coordinates the connection on f^*TY is $\Gamma^\alpha_{\beta\gamma}(f) \frac{\partial f^\gamma}{\partial x^i}$ for x-derivatives and $\Gamma^\alpha_{\beta\gamma}(f) \frac{\partial f^\gamma}{\partial t}$ for t-derivatives. Therefore in local coordinates

$$\nabla\nabla f = \frac{\partial^2 f^\alpha}{\partial x^i \partial x^j} - \frac{\partial f^\alpha}{\partial x^k} \Gamma^k_{ij} + \Gamma^\alpha_{\beta\gamma}(f) \frac{\partial f^\beta}{\partial x^i} \frac{\partial f^\gamma}{\partial x^j}$$

and as before

$$\Delta f = \nabla_v \nabla_v f = g^{ij}\left\{\frac{\partial^2 f^\alpha}{\partial x^i \partial x^j} - \frac{\partial f^\alpha}{\partial x^k} \Gamma^k_{ij} + \Gamma^\alpha_{\beta\gamma}(f) \frac{\partial f^\alpha}{\partial x^i} \frac{\partial f^\gamma}{\partial x^j}\right\}.$$

Note that f itself is not a section of a vector bundle, but $\nabla\nabla f$ is defined in such a way as to be symmetric. If g is a section of a general Riemannian connected vector bundle V the second covariant derivative $\nabla\nabla g$ is not in general symmetric. The asymmetry is given by the curvature tensor R of the bundle V. Thus

$$\nabla_w \nabla_v g = \nabla_v \nabla_w g + R(v,w)g$$

where $R(v,w)$ is anti-symmetric in v and w and has values in $L(V,V)$; thus $R \in L_A^2(TX, L(V,V))$. If $V = TX$ this is the Riemannian curvature $R(v,w)$. The Ricci curvature $R \in L(TX,TX)$ is defined by contraction. Then

$$R(v,w)\,\nabla_v = R\,\nabla_w.$$

<u>Theorem.</u> Let χ be the potential energy density

$$\chi = \tfrac{1}{2}|\nabla f|^2 = \tfrac{1}{2}\langle \nabla_w f, \nabla_w f \rangle$$

$$\chi = \frac{1}{2} g^{ij} \frac{\partial f^\alpha}{\partial x^i} \frac{\partial f^\beta}{\partial x^j} h_{\alpha\beta}(f).$$

Let R_X be the Ricci curvature on X and $R_Y(f)(v,w)$ the Riemannian curvature on f^*TY. Then

$$\frac{\partial\chi}{\partial t} = \Delta\chi - |\nabla\nabla f|^2 - \langle R_X \nabla_v f, \nabla_v f \rangle$$

$$+ \langle R_Y(f)(\nabla_v f, \nabla_w f)\nabla_v f, \nabla_w f \rangle.$$

At the origin in geodesic coordinates

$$\frac{\partial\chi}{\partial t} = g^{ij} \frac{\partial^2\chi}{\partial x^i \partial x^j} - g^{ik} g^{j\ell} \frac{\partial^2 f^\alpha}{\partial x^i \partial x^j} \frac{\partial^2 f^\beta}{\partial x^k \partial x^\ell} h_{\alpha\beta}$$

$$- R^{ij} \frac{\partial f^\alpha}{\partial x^i} \frac{\partial f^\beta}{\partial x^j} h_{\alpha\beta} + g^{ik} g^{j\ell} R_{\alpha\beta\gamma\delta} \frac{\partial f^\alpha}{\partial x^i} \frac{\partial f^\beta}{\partial x^j} \frac{\partial f^\gamma}{\partial x^k} \frac{\partial f^\delta}{\partial x^\ell}.$$

<u>Corollary.</u> If $R_Y \leq 0$ and $R_X \geq -C$

$$\frac{\partial\chi}{\partial t} \leq \Delta\chi + C\chi - |\nabla\nabla f|^2.$$

<u>Proof.</u> We have

$$\frac{\partial\chi}{\partial t} = \langle \frac{\partial}{\partial t} \nabla_w f, \nabla_w f \rangle = \langle \nabla_w \frac{\partial f}{\partial t}, \nabla_w f \rangle$$

since for a Riemannian connection the covariant derivative of the inner product is zero. In local coordinates we must remember $h_{\alpha\beta}(f)$ depends on f and hence also on t. This introduces terms

$\frac{\partial h_{\alpha\beta}}{\partial y^\delta}(f)\ \frac{\partial f^\delta}{\partial t}$. Using symmetry we can convert the $\frac{\partial h_{\alpha\beta}}{\partial y^\delta}(f)$ into a Christoffel symbol $\Gamma^\alpha_{\beta\gamma}(f)$. We get

$$\frac{\partial X}{\partial t} = g^{ij}\left[\frac{\partial^2 f^\alpha}{\partial x^i \partial t} + \Gamma^\alpha_{\gamma\delta}(f)\ \frac{\partial f^\gamma}{\partial x^i}\ \frac{\partial f^\delta}{\partial t}\right]\frac{\partial f^\beta}{\partial x^j}\ h_{\alpha\beta}(f).$$

Note that the quantity in brackets is $\nabla\frac{\partial}{\partial t}f$, using the connection $\Gamma^\alpha_{\gamma\delta}(f)\ \frac{\partial f^\gamma}{\partial x^i}$ on f^*TY. Likewise

$$\nabla X = \langle\ \nabla\nabla_w f,\ \nabla_w f\ \rangle$$

$$\frac{\partial X}{\partial x^k} = g^{ij}\left[\frac{\partial^2 f^\alpha}{\partial x^k \partial x^i} - \frac{\partial f^\alpha}{\partial x^\ell}\Gamma^\ell_{ki} + \Gamma^\alpha_{\gamma\delta}\ \frac{\partial f^\gamma}{\partial x^k}\ \frac{\partial f^\delta}{\partial x^i}\right]\frac{\partial f^\beta}{\partial x^j}\ h_{\alpha\beta}(f).$$

Again the quantity in brackets is $\nabla\nabla f$, using the connection on $L(TX,\ f^*TY)$. Then

$$\nabla\nabla X = \langle\ \nabla\nabla\nabla_w f,\ \nabla_w f\ \rangle + \langle\ \nabla\nabla_w f,\ \nabla\nabla_w f\ \rangle$$

$$\Delta X = \nabla_v\nabla_v X = \langle\ \nabla_v\nabla_v\nabla_w f,\ \nabla_w f\ \rangle + \langle\ \nabla_v\nabla_w f,\ \nabla_v\nabla_w f\ \rangle.$$

At the origin of geodesic coordinates the $\Gamma^\alpha_{\beta\gamma}$ and Γ^i_{jk} are zero, and differences of their derivatives express curvature tensors. We get

$$\begin{aligned}\Delta X &= g^{km}\ \frac{\partial^2 X}{\partial x^k \partial x^m}\\ &= g^{ij}g^{km}\left[\frac{\partial^3 f^\alpha}{\partial x^i \partial x^k \partial x^m} - \frac{\partial f^\alpha}{\partial x^\ell}\ \frac{\partial \Gamma^\ell_{ik}}{\partial x^m} + \frac{\partial \Gamma^\alpha_{\gamma\delta}}{\partial y^\epsilon}\ \frac{\partial f^\epsilon}{\partial x^m}\ \frac{\partial f^\gamma}{\partial x^i}\ \frac{\partial f^\delta}{\partial x^k}\right]\frac{\partial f^\beta}{\partial x^j}\ h_{\alpha\beta}\\ &\qquad + g^{ij}g^{km}\frac{\partial^2 f^\alpha}{\partial x^i \partial x^k}\ \frac{\partial^2 f^\beta}{\partial x^j \partial x^m}\ h^{\alpha\beta}.\end{aligned}$$

Since f satisfies the heat equation we can substitute Δf for $\frac{\partial f}{\partial t}$ in the formula for $\frac{\partial X}{\partial t}$. We get

$$\frac{\partial \chi}{\partial t} = \langle \nabla_w \frac{\partial f}{\partial t}, \nabla_w f \rangle = \langle \nabla_w \nabla_v \nabla_v f, \nabla_w f \rangle.$$

In geodesic coordinates

$$\frac{\partial \chi}{\partial t} = g^{ij} g^{km} \left[\frac{\partial^3 f^\alpha}{\partial x^i \partial x^k \partial x^m} - \frac{\partial f^\alpha}{\partial x^\ell} \frac{\partial \Gamma^\ell_{km}}{\partial x^i} \right.$$

$$\left. + \frac{\partial \Gamma^\alpha_{\gamma\epsilon}}{\partial y^\delta} \frac{\partial f^\delta}{\partial x^i} \frac{\partial f^\gamma}{\partial x^k} \frac{\partial f^\epsilon}{\partial x^m} \right] \frac{\partial f^\beta}{\partial x^j} h_{\alpha\beta}.$$

Now $\nabla_w \nabla_v (\nabla_v f) = \nabla_v \nabla_w (\nabla_v f) + R(v,w)\nabla_v f$ where $R(v,w)$ is the curvature on the bundle $L(TX, f^*TY)$ of which $\nabla_v f$ is a section. This is minus the curvature of TX plus the curvature of f^*TY. If $R_X(v,w)$ is the Riemannian curvature on X then $R_X(v,w)\nabla_v f = R_X \nabla_w f$ where R_X is the Ricci curvature. If $R_Y(\ ,\ )$ is the Riemannian curvature on Y then $R_Y(f)(\nabla_v f, \nabla_w f)$ is the curvature of the pull-back bundle f^*TY. Thus

$$R(v,w)\nabla_v f = - R_X \nabla_w f + R_Y(f)(\nabla_v f, \nabla_w f)\nabla_v f.$$

Since $\nabla\nabla f$ is symmetric, $\nabla_v \nabla_w f = \nabla_w \nabla_v f$ and $\nabla_v \nabla_w \nabla_v f = \nabla_v \nabla_v \nabla_w f$. Therefore

$$\frac{\partial \chi}{\partial t} = \Delta \chi - |\nabla\nabla f|^2 - \langle R_X \nabla_w f, \nabla_w f \rangle$$

$$+ \langle R_Y(f)(\nabla_v f, \nabla_w f)\nabla_v f, \nabla_w f \rangle.$$

In geodesic coordinates the $\frac{\partial \Gamma^\ell_{km}}{\partial x^i}$ and $\frac{\partial \Gamma^\alpha_{\gamma\delta}}{\partial y^\epsilon}$ subtract to give the curvature terms in the formula given in the statement of the theorem. Since this formula is in terms of tensors it is invariant under coordinate change and hence valid in a general coordinate system.

The important observation is that if $R_Y \leq 0$ the highest order non-linear terms involving $\nabla\nabla f \cdot \nabla\nabla f$ or $\nabla f \cdot \nabla f \cdot \nabla f \cdot \nabla f$ enter negatively, while terms $\nabla\nabla f \cdot \nabla f \cdot \nabla f$ or $\nabla\nabla\nabla f \cdot \nabla f$ are absent. We could handle terms $\nabla\nabla f \cdot \nabla f$ or $\nabla f \cdot \nabla f \cdot \nabla f$ because they are lower order, but luckily they are absent also.

3. Next we wish to derive an equation for the normal derivative $\frac{\partial X}{\partial \nu}$.

Theorem. There exists a second order linear operator $a\nabla\nabla f + b\nabla f$ with smooth coefficients depending only on $f|\partial X \times [0,\omega) = h$ such that if m is the mean curvature of ∂X then on $\partial X \times [0,\omega)$

$$\frac{\partial X}{\partial \nu} - 2mX = a\nabla\nabla f + b\nabla f.$$

Proof. We use the fact that f equals the known function h on $\partial X \times [0,\omega)$ and try to express as much as possible in terms of h. Let $v,w,\ldots$ again denote general tangent vectors but let $\pi,\rho,\ldots$ denote vectors tangent to ∂X and ν the normal vector. If a vector v is repeated we trace over all of TX but if a vector π is repeated we trace only over $T\partial X$. In local coordinates choose a chart $\{x^1,\ldots,x^n\}$ such that x^n is the distance to ∂X. Then $\nabla_\nu = \frac{\partial}{\partial x^n}$. Let $i,j,k,\ldots$ be indices ranging $1 \leq i \leq n$ and let $p,q,r,\ldots$ be restricted to $1 \leq p \leq n-1$. We denote the covariant derivative along ∂X by ∇^∂. Then $\nabla^\partial f = \nabla^\partial h$ and $\nabla^\partial\nabla^\partial f = \nabla^\partial\nabla^\partial f$ are known in terms of h alone. If π is tangent to ∂X then $\nabla_\pi f = \nabla^\partial_\pi f = \nabla^\partial_\pi h$. For second derivatives there is a correction given by the second fundamental form $F(\pi,\rho)$ of ∂X.

Lemma. $\nabla_\pi\nabla_\rho f = \nabla^\partial_\pi \nabla^\partial_\rho f - F(\pi,\rho)\, \nabla_\nu f$.

Proof. In local coordinates with $1 \leq p,\ q \leq n-1$

$$\nabla\nabla f = \frac{\partial^2 f^\alpha}{\partial x^p \partial x^q} - \frac{\partial f^\alpha}{\partial x^k}\Gamma^k_{pq} + \Gamma^\alpha_{\beta\gamma}(f)\,\frac{\partial f^\beta}{\partial x^i}\,\frac{\partial f^\gamma}{\partial x^j}\,.$$

In the corresponding term of $\nabla^\partial\nabla^\partial f$ the summation on k is restricted to $1 \leq k \leq n-1$, omitting the term $\frac{\partial f^\alpha}{\partial x^n}\Gamma^n_{pq}$. Now Γ^n_{pq} is the matrix of the second fundamental form $F(\pi,\rho)$ and $\frac{\partial f^\alpha}{\partial x^n}$ is the normal derivative $\nabla_\nu f$. Thus we have shown the formula.

Taking traces,

$$\Delta f = \nabla_v \nabla_v f = \nabla_\pi \nabla_\pi f + \nabla_\nu \nabla_\nu f$$

$$\nabla_\pi \nabla_\pi f = \nabla^\partial_\pi \nabla^\partial_\pi f + F(\pi,\pi)\ \nabla_\nu f$$

$$\Delta^\partial f = \nabla^\partial_\pi \nabla^\partial_\pi f.$$

$F(\pi,\pi) = m$ the mean curvature. Thus we have the

Corollary. On ∂X

$$\Delta f = \Delta^\partial f + \nabla_\nu \nabla_\nu f - m\ \nabla_\nu f.$$

Since f satisfies the heat equation, $\frac{\partial f}{\partial t} = \Delta f$. But $f = h$ on $\partial X \times [0,\omega)$ and h is independent of t. Thus $\Delta f = 0$ on $\partial X \times [0,\omega)$. Then $\nabla_\nu \nabla_\nu f = m\ \nabla_\nu f - \Delta^\partial f$. Now

$$X = \frac{1}{2}\langle \nabla_\nu f,\ \nabla_\nu f\rangle$$

$$X = \frac{1}{2}\langle \nabla_\pi f,\ \nabla_\pi f\rangle + \frac{1}{2}\langle \nabla_\nu f,\ \nabla_\nu f\rangle$$

$$\nabla_\nu X = \langle \nabla_\nu \nabla_\pi f,\ \nabla_\pi f\rangle + \langle \nabla_\nu \nabla_\nu f,\ \nabla_\nu f\rangle$$

$$\nabla_\nu X = \langle \nabla_\nu \nabla_\pi f,\ \nabla_\pi f\rangle + m\,\langle \nabla_\nu f, \nabla_\nu f\rangle - \langle \Delta^\partial f, \nabla_\nu f\rangle.$$

In local coordinates

$$\frac{\partial X}{\partial x^n} = g^{pq}\left[\frac{\partial^2 f^\alpha}{\partial x^n \partial x^p} - \frac{\partial f^\alpha}{\partial x^i}\Gamma^i_{np} + \Gamma^\alpha_{\gamma\delta}\frac{\partial f^\gamma}{\partial x^n}\frac{\partial f^\delta}{\partial x^p}\right]\frac{\partial f^\beta}{\partial x^q}\ h_{\alpha\beta}$$

$$+ g^{pq}\,\Gamma^n_{pq}\frac{\partial f^\alpha}{\partial x^n}\frac{\partial f^\beta}{\partial x^n}h_{\alpha\beta}$$

$$- g^{pq}\left[\frac{\partial^2 f^\alpha}{\partial x^p \partial x^q} - \frac{\partial f^\alpha}{\partial x^r}\Gamma^r_{np} + \Gamma^\alpha_{\gamma\delta}\frac{\partial f^\gamma}{\partial x^p}\frac{\partial f^\delta}{\partial x^q}\right]\frac{\partial f^\beta}{\partial x^n}\ h_{\alpha\beta}.$$

The term

$$m\langle \nabla_\nu f, \nabla_\nu f\rangle = 2mX - m\langle \nabla_\pi f, \nabla_\pi f\rangle.$$

Thus

$$\nabla_\nu X - 2mX = \langle \nabla_\nu \nabla_\pi f, \nabla_\pi f\rangle - m\langle \nabla_\pi f, \nabla_\pi f\rangle - \langle \nabla_\nu f, \Delta^\partial f\rangle.$$

Now $\nabla_\pi f$ and $\Delta^\partial f$ are known on $\partial X \times [0,\omega)$ in terms of h. Therefore the right hand side may be regarded as a second order linear differential operator of the form $a\nabla\nabla f + b\nabla f$ with smooth coefficients depending on h, since in each term all but one of the f-derivatives may be expressed as h-derivatives. In local coordinates it has the form

$$g^{pq}\left[\frac{\partial^2 f^\alpha}{\partial x^n \partial x^p} - \frac{\partial f^\alpha}{\partial x^i}\Gamma^i_{np} + \Gamma^\alpha_{\gamma\delta}\frac{\partial f^\gamma}{\partial x^n}\frac{\partial h^\delta}{\partial x^p}\right]\frac{\partial h^\beta}{\partial x^q}h_{\alpha\beta}$$

$$-g^{pq}\,\Gamma^n_{pq}\,g^{rs}\frac{\partial f^\alpha}{\partial x^r}\frac{\partial h^\beta}{\partial x^s}h_{\alpha\beta}$$

$$-g^{pq}\left[\frac{\partial^2 h^\alpha}{\partial x^p \partial x^q} - \frac{\partial h^\alpha}{\partial x^r}\Gamma^r_{np} + \Gamma^\alpha_{\gamma\delta}\frac{\partial h^\gamma}{\partial x^p}\frac{\partial h^\delta}{\partial x^q}\right]\frac{\partial f^\beta}{\partial x^n}h_{\alpha\beta}.$$

The important observation is that there are no third order terms, i.e. products $\nabla_\nu\nabla_\nu f\cdot\nabla_\nu f$ or $\nabla_\nu f\cdot\nabla_\nu f\cdot\nabla_\nu f$.

4. We can make a similar calculation for the kinetic energy

$$\varkappa = \frac{1}{2}\left|\frac{\partial f}{\partial t}\right|^2 = \frac{1}{2}\left\langle\frac{\partial f}{\partial t}, \frac{\partial f}{\partial t}\right\rangle$$

$$\varkappa = \frac{1}{2}\frac{\partial f^\alpha}{\partial t}\frac{\partial f^\beta}{\partial t}h_{\alpha\beta}(f)$$

$$\frac{\partial\varkappa}{\partial t} = \left\langle\frac{\partial^2 f}{\partial t^2}, \frac{\partial f}{\partial t}\right\rangle$$

$$\nabla\varkappa = \left\langle\nabla\frac{\partial f}{\partial t}, \frac{\partial f}{\partial t}\right\rangle$$

$$\nabla\nabla\varkappa = \left\langle\nabla\nabla\frac{\partial f}{\partial t}, \frac{\partial f}{\partial t}\right\rangle + \left\langle\nabla\frac{\partial f}{\partial t}, \nabla\frac{\partial f}{\partial t}\right\rangle$$

$$\Delta\varkappa = \nabla_v\nabla_v\varkappa = \left\langle\nabla_v\nabla_v\frac{\partial f}{\partial t}, \frac{\partial f}{\partial t}\right\rangle + \left\langle\nabla_v\frac{\partial f}{\partial t}, \nabla_v\frac{\partial f}{\partial t}\right\rangle.$$

Since f satisfies the heat equation

$$\frac{\partial f}{\partial t} = \Delta f = \nabla_v\nabla_v f$$

$$\frac{\partial^2 f}{\partial t^2} = \frac{\partial}{\partial t}\nabla_v\nabla_v f.$$

Now $\nabla_v f$ is a section of the bundle $L(TX, f^*TY)$. This bundle is not flat in the t direction since f^*TY varies with t. In the t-direction the connection is given by $\Gamma^{\alpha}_{\beta\gamma}(f)\frac{\partial f^{\gamma}}{\partial t}$. Therefore ∇_v and $\frac{\partial}{\partial t}$ do not commute on f^*TY. Their commutator is given by the curvature $R_Y(f)(\nabla_v f, \frac{\partial f}{\partial t})$ where $R_Y(\ ,\ )$ is the Riemannian curvature on Y. The operators ∇_v and $\frac{\partial}{\partial t}$ do commute on TX, so there are no curvature terms here. Thus

$$\frac{\partial}{\partial t}\nabla_v\nabla_v f = \nabla_v\frac{\partial}{\partial t}\nabla_v f + R_Y(f)(\nabla_v f, \frac{\partial f}{\partial t})\nabla_v f.$$

We do have $\frac{\partial}{\partial t}\nabla_v f = \nabla_v \frac{\partial}{\partial t} f$, so

$$\nabla_v\frac{\partial}{\partial t}\nabla_v f = \nabla_v\nabla_v\frac{\partial}{\partial t} f.$$

Also $\langle\nabla_v \frac{\partial f}{\partial t}, \nabla_v \frac{\partial f}{\partial t}\rangle = |\nabla \frac{\partial f}{\partial t}|^2$. Thus we have proved the following:

<u>Theorem.</u>

$$\frac{\partial \varkappa}{\partial t} = \Delta\varkappa - |\nabla \frac{\partial f}{\partial t}|^2 + \langle R_Y(f)(\nabla_v f, \frac{\partial f}{\partial t})\nabla_v f, \frac{\partial f}{\partial t}\rangle.$$

<u>Corollary.</u> If $R_Y \leq 0$ then

$$\frac{\partial\varkappa}{\partial t} \leq \Delta\varkappa - |\nabla \frac{\partial f}{\partial t}|^2.$$

In local coordinates we have

$$\frac{\partial\varkappa}{\partial t} = g^{ij}\left[\frac{\partial^2\varkappa}{\partial x^i\partial x^j} - \frac{\partial\varkappa}{\partial x^k}\Gamma^k_{ij}\right] - g^{ij}\frac{\partial^2 f^{\alpha}}{\partial x^i\partial t}\frac{\partial^2 f^{\beta}}{\partial x^j\partial t}h_{\alpha\beta}$$

$$+ g^{ij}R_{\alpha\beta\gamma\delta}\frac{\partial f^{\alpha}}{\partial x^i}\frac{\partial f^{\beta}}{\partial t}\frac{\partial f^{\gamma}}{\partial x^j}\frac{\partial f^{\delta}}{\partial t}.$$

<u>Theorem.</u> On $\partial X \times [0,\omega)$ we have $\varkappa = 0$ and $\frac{\partial\varkappa}{\partial \nu} = 0$.

<u>Proof.</u>

$$\varkappa = \frac{1}{2}\langle\frac{\partial f}{\partial t}, \frac{\partial f}{\partial t}\rangle$$

$$\nabla_\nu\varkappa = \langle\nabla_\nu\frac{\partial f}{\partial t}, \frac{\partial f}{\partial t}\rangle.$$

But on $\partial X \times [0,\omega)$ we have $f = h$ and h is independent of t, so $\frac{\partial f}{\partial t} = 0$ on $\partial X \times [0,\omega)$.

5. We define the total potential energy of the map $f:X \to Y$ to be

$$E(f) = \int_X \chi = \frac{1}{2} \int_X \langle \nabla_W f, \nabla_W f \rangle .$$

We compute its derivative or first variation. An infinitesimal variation of the map f is a section k of the bundle $f^* TY$.

Theorem.

$$DE(f)(k) = - \int_X \langle \Delta f , k \rangle + \int_{\partial X} \langle \nabla_\nu f, k \rangle .$$

Proof.

$$E(f) = \frac{1}{2} \int_X \langle \nabla_W f, \nabla_W f \rangle$$

$$DE(f)(k) = \int_X \langle \nabla_W f, \nabla_W k \rangle .$$

By Green's theorem

$$\int_X \langle \nabla_W f , \nabla_W k \rangle + \int_X \langle \nabla_W \nabla_W f, k \rangle = \int_{\partial X} \langle \nabla_\nu f, k \rangle .$$

Since $\nabla_W \nabla_W f = \Delta f$ the result follows. If the variation $k = 0$ on ∂X then the boundary term is zero. If $f:X \times [0,\omega) \to Y$ we may regard $E(f)$ as a function of t.

We also define the total kinetic energy

$$K(f) = \int_X \kappa = \frac{1}{2} \int_X \left| \frac{\partial f}{\partial t} \right|^2 .$$

It is also a function of t.

Theorem.

$$\frac{d}{dt} E(f) = -2K(f).$$

Proof.

$$\frac{d}{dt} E(f) = DE(f)\left(\frac{\partial f}{\partial t}\right) = - \int_X \left\langle \Delta f, \frac{\partial f}{\partial t} \right\rangle$$

$$= \int_X \left\langle \frac{\partial f}{\partial t}, \frac{\partial f}{\partial t} \right\rangle = -2K(f)$$

since $\frac{\partial f}{\partial t} = 0$ on $\partial X \times [0,\omega)$

Theorem. $\frac{d}{dt} K(f) \leq 0$.

Proof.

$$\frac{d}{dt} K(f) = \frac{d}{dt} \int_X \varkappa = \int_X \frac{\partial \varkappa}{\partial t} \leq \int_X \Delta\varkappa$$

since $\frac{\partial \varkappa}{\partial t} \leq \Delta\varkappa$. Also $\frac{\partial \varkappa}{\partial \nu} = 0$ on $\partial X \times [0,\omega)$ so by Green's theorem

$$\int_X \Delta\varkappa = \int_{\partial X} \frac{\partial}{\partial \nu} = 0.$$

Therefore $\frac{d}{dt} K(f) \leq 0$.

Corollary. $E(f)$ is positive, monotonically decreasing and convex.

Proof. Clearly $E(f) \geq 0$ and $K(f) \geq 0$. Then $\frac{d}{dt} E(f) = -2K(f) \leq 0$ and $\frac{d^2}{dt^2} E(f) = -2\frac{d}{dt} K(f) \geq 0$.

Corollary. $K(f)$ is positive and monotonically decreasing. If $\omega = \infty$ then $K(f) \to 0$ as $t \to \infty$.

Proof. We know $K(f)$ is positive, and monotonically decreasing since $\frac{d}{dt} K(f) \leq 0$. Unless $K(f) \to 0$ as $t \to \infty$ the quantity $E(f)$ would become negative.

Corollary. With a constant C independent of τ

$$\int_{X\times\tau} |\nabla f|^2 \leq C$$

$$\int_{X\times\tau} \left|\frac{\partial f}{\partial t}\right|^2 \leq C.$$

6. These estimates came from integrating the inequality for κ. If we integrate the inequality for χ we get the following:

Theorem. With a constant C independent of τ,

$$\int_{X\times\tau} |\nabla\nabla f|^2 \leq C\left\{\int_{\partial X\times\tau} |\nabla\nabla f| + \int_{\partial X\times\tau} |\nabla f|^2 + 1\right\}.$$

Proof.

$$\frac{d}{dt} E(f) = \frac{d}{dt}\int_{X\times\tau} \chi = \int_{X\times\tau} \frac{\partial\chi}{\partial t} \leq \int_{X\times\tau} \Delta\chi + C\int_{X\times\tau} \chi - \int_{X\times\tau} |\nabla\nabla f|^2.$$

Using Green's Theorem

$$\int_{X\times\tau} \Delta\chi = \int_{\partial X\times\tau} \frac{\partial\chi}{\partial\nu} = \int_{\partial X\times\tau} 2m\,\chi + a\nabla\nabla f + b\nabla f.$$

Since $\chi = \frac{1}{2}|\nabla f|^2$,

$$\int_{X\times\tau} \Delta\chi \leq C\left\{\int_{\partial X\times\tau} |\nabla\nabla f| + \int_{\partial X\times\tau} |\nabla f|^2 + 1\right\}.$$

Also $\int_{X\times\tau} \chi \leq C$. Now $\frac{d}{dt} E(f) = -2K(f)$ and $K(f) \leq C$ so $\frac{d}{dt} E(f) \geq -2C$. Therefore with a constant C independent of τ,

$$\int_{X\times\tau} |\nabla\nabla f|^2 \leq C\left\{\int_{\partial X\times\tau} |\nabla\nabla f| + \int_{\partial X\times\tau} |\nabla f|^2 + 1\right\}.$$

We can improve this result as follows. Regarding $X \times \tau$ as a manifold we have defined $\|f\|_{L^2_2(X\times\tau)}$ as

$$\| f \|^2_{L^2_2(X\times\tau)} = \int_{X\times\tau} |\nabla\nabla f|^2 + \int_{X\times\tau} |\nabla f|^2 + \int_{X\times\tau} |f|^2$$

where the last term is automatically bounded since f is.

Theorem. With a constant C independent of τ,

$$\| f \|^2_{L^2_2(X\times\tau)} \leq C\left\{\int_{\partial X\times\tau} |\nabla\nabla f| + 1\right\} .$$

Proof.

$$\int_{\partial X\times\tau} |\nabla f|^2 = \| \nabla f \|^2_{L^2(\partial X\times\tau)} \leq \| \nabla f \|^2_{L^2_{2/3}(X\times\tau)}$$

since $2/3 > 1/2$. Now

$$\| \nabla f \|_{L^2_{2/3}(X\times\tau)} \leq C \| f \|_{L^2_{5/3}(X\times\tau)}$$

and by interpolation

$$\| f \|_{L^2_{5/3}(X\times\tau)} \leq C \| f \|^{2/3}_{L^2_2(X\times\tau)} \| f \|^{1/3}_{L^2_1(X\times\tau)} .$$

But

$$\| f \|^2_{L^2_1(X\times\tau)} = \int_{X\times\tau} |\nabla f|^2 + \int_{X\times\tau} |f|^2 \leq C$$

independent of τ. Therefore

$$\int_{\partial X\times\tau} |\nabla f|^2 \leq C \| f \|^{4/3}_{L^2_2(X\times\tau)} .$$

Using the previous theorem

$$\| f \|^2_{L^2_2(X\times\tau)} \leq C\left\{\int_{\partial X\times\tau} |\nabla\nabla f| + \int_{\partial X\times\tau} |\nabla f|^2 + 1\right\}$$

$$\leq C\left\{\int_{\partial X\times\tau} |\nabla\nabla f| + 1\right\} + C \| f \|^{4/3}_{L^2_2(X\times\tau)} .$$

Since $4/3 < 2$, this proves the result.

Now suppose $0 \leq \tau < \omega - \delta$ where δ is fixed with $\delta = 1$ if $\omega = \infty$ and $\delta < \omega/4$ otherwise. In the following C denotes various constants independent of τ, but there may be a dependence on δ.

This will not concern us, since we use only one value of δ. We can integrate the previous results over $[\tau,\tau+\delta]$. This gives us

$$\int_{X\times[\tau,\tau+\delta]} |\nabla f|^2 \leq C$$

$$\int_{X\times[\tau,\tau+\delta]} \left|\frac{\partial f}{\partial t}\right|^2 \leq C$$

$$\int_{X\times[\tau,\tau+\delta]} |\nabla\nabla f|^2 \leq C\left\{\int_{\partial X\times[\tau,\tau+\delta]} |\nabla\nabla f| + 1\right\}.$$

Recalling that

$$\|f\|_{L_2^2(X\times[\tau,\tau+\delta])} = \int_{X\times[\tau,\tau+\delta]} |\nabla\nabla f|^2 + \int_{X\times[\tau,\tau+\delta]} \left|\frac{\partial f}{\partial t}\right|^2 + \int_{X\times[\tau,\tau+\delta]} |\nabla f|^2 + \int_{X\times[\tau,\tau+\delta]} |f|^2$$

we have proved the following:

Theorem. With a constant C independent of τ,

$$\|f\|^2_{L_2^2(X\times[\tau,\tau+\delta])} \leq C\left\{\int_{\partial X\times[\tau,\tau+\delta]} |\nabla\nabla f| + 1\right\}.$$

7. We prove some preparatory lemmas.

Lemma (a). If $0 < \alpha < 1$ and $x,y \geq 0$, $(x+y)^\alpha \leq x^\alpha + y^\alpha$.

Proof. Since both sides are homogeneous we can restrict our attention to the quarter circle $x^2+y^2 = 1$, $x \geq 0$, $y \geq 0$. By Lagrange multipliers the maximum of $(x+y)^\alpha/(x^\alpha+y^\alpha)$ occurs at an endpoint $x = 1$, $y = 0$ or $x = 0$, $y = 1$ when $0 < \alpha < 1$ (and in the middle $x = y$ for $\alpha > 1$).

Lemma (b). Suppose $0 < \alpha < 1$ and $x,y \geq 0$. For any constant C there is a constant C' depending only on C such that

$$x \leq C(x+y)^\alpha \text{ implies } x \leq C'(1+y)^\alpha.$$

<u>Proof</u>. Choose $C' = (2C)^{1/(1-\alpha)} + 2C$.

<u>Case I</u>. If $x - Cx^{\alpha} \leq x/2$ then $x/2 \leq Cx^{\alpha}$ so

$$x \leq (2C)^{1/(1-\alpha)} \leq C' \leq C'(1+y)^{\alpha}$$

<u>Case II</u>. If $x/2 \leq x - Cx^{\alpha}$ we know from Lemma (a) that $x \leq Cx^{\alpha} + Cy^{\alpha}$ so $x/2 \leq x - Cx^{\alpha} \leq Cy^{\alpha}$. Then

$$x \leq 2Cy^{\alpha} \leq C'(1+y)^{\alpha}.$$

<u>Lemma (c)</u>. Suppose $0 < \alpha < 1$ and C is a constant. Let c_k be a sequence of terms with $c_k \geq 0$ such that

$$c_{k+1} \leq C(1+c_k)^{\alpha}.$$

Then the sequence c_k is bounded.

<u>Proof</u>. By L'Hopital's rule

$$\lim_{x\to\infty} \frac{(1+x)^{\alpha}}{x} = \lim_{x\to\infty} \alpha(1+x)^{\alpha-1} = 0.$$

Therefore we can choose X such that if $x \geq X$ then $x \geq C(1+x)^{\alpha}$. In particular $X \geq C(1+X)^{\alpha}$. There are two possibilities.

<u>Case I</u>. All $c_k \geq X$.
Then $c_{k+1} \leq C(1+c_k)^{\alpha} \leq c_k$, so the sequence is monotone decreasing and hence is bounded.

<u>Case II</u>. Some $c_k \leq X$. Then $c_{k+1} \leq C(1+c_k)^{\alpha} \leq C(1+X)^{\alpha} \leq X$ so $c_k \leq X$ implies $c_{k+1} \leq X$ and the sequence is again bounded.

8. **Theorem.** There is a constant C independent of τ such that for $0 \leq \tau < \omega - \delta$ (or $0 \leq \tau < \infty$ if $\omega = \infty$)

$$\| f \|_{L_2^2(X \times [\tau, \tau + \delta])} \leq C.$$

Proof. From Section 6 we know that

$$\| f \|^2_{L_2^2(X \times [\tau, \tau+\delta])} \leq C \left\{ \int_{\partial X \times [\tau, \tau+\delta]} |\nabla\nabla f| + 1 \right\}$$

We can estimate the boundary integral

$$\int_{\partial X \times [\tau, \tau+\delta]} |\nabla\nabla f| \leq C \| \nabla\nabla f \|_{L^1(\partial X \times [\tau, \tau+\delta])}$$

$$\leq C \| \nabla\nabla f \|_{L^{4/3}(\partial X \times [\tau, \tau+\delta])}$$

where $4/3$ is a convenient choice of $p > 1$. Since $3/4 < 4/5$ we have

$$\| \nabla\nabla f \|_{L^{4/3}(\partial X \times [\tau, \tau+\delta])} \leq C \| \nabla\nabla f \|_{L^{4/3}_{4/5}(X \times [\tau, \tau+\delta])}$$

$$\| \nabla\nabla f \|_{L^{4/3}_{4/5}(X \times [\tau, \tau+\delta])} \leq C \| f \|_{L^{4/3}_{14/5}(X \times [\tau, \tau+\delta])}$$

(We remind the reader that the power L^p is on the top and the fractional number of derivatives n is on the bottom in L_n^p).

Next we apply Gårding's inequality to the equation $\frac{\partial f}{\partial t} - \Delta f$ with boundary condition $f | \partial X \times [\alpha, \omega]$. Although f is a system we can estimate each f^α separately. We know that

$$\| f \|_{L_n^p(X \times [\tau, \tau+\delta])} \leq C \Big(\| \frac{\partial f}{\partial t} - \Delta f \|_{L^p_{n-2}(X \times [\tau-\delta, \tau+\delta])}$$

$$+ \| f | \partial X \times [\tau-\delta, \tau+\delta] \|_{\partial L_n^p(X \times [\tau-\delta, \tau+\delta])} + \| f \|_{L_0^p(X \times [\tau-\delta, \tau+\delta])} \Big)$$

with a constant C independent of f. Since the equation is invariant under translation in the t direction we also see that C is independent of τ. Now $f|\partial X \times [\tau-\delta,\tau+\delta]$ is equal to the given function h which is independent of t so

$$\| f|\partial X \times [\tau-\delta,\tau+\delta] \|_{\partial L^p_n(X\times[\tau-\delta,\tau+\delta])} \leq C$$

independent of τ. Also f is continuous and bounded so

$$\| f \|_{L^p_0(X\times[\tau-\delta,\tau+\delta])} \leq C$$

independent of τ. Since f satisfies the heat equation

$$\frac{\partial f}{\partial t} = \Delta f = \Delta f + \Gamma(f)\ \nabla f^2$$

we can conclude that

$$\| f \|_{L^p_n(X\times[\tau,\tau+\delta])} \leq C(1+\| \Gamma(f)\nabla f^2 \|_{L^p_{n-2}(X\times[\tau-\delta,\tau+\delta])})$$

In the case at hand we have $p' = 4/3$ and $n = 14/5$. Thus

$$\| f \|_{L^{4/3}_{14/5}(X\times[\tau,\tau+\delta])} \leq C(1+\| \Gamma(f)\nabla f^2 \|_{L^{4/3}_{4/5}(X\times[\tau-\delta,\tau+\delta])}).$$

Since $\Gamma(f)\nabla f^2$ is a polynomial differential operator of type $(2,1)$ we can apply the theorem of Part IV Section 6. In that theorem let $n = 2$, $k = 1$, $p = 4/3$, $q = 2$, $r = 4/5$ and $s = 2$. Then $4/5 + 1 < 2$ and $(4/3)(4/5 + 2) < 2\cdot 2$ so the hypotheses are satisfied. Thus

$$\| \Gamma(f)\nabla f^2 \|_{L^{4/3}_{4/5}(X\times[\tau-\delta,\tau+\delta])} \leq C(1 + \| f \|_{L^2_2(X\times[\tau-\delta,\tau+\delta])})^{3/2}.$$

Combining these results

$$\| f \|^2_{L^2_2(X\times[\tau,\tau+\delta])} \leq C(1 + \| f \|_{L^2_2(X\times[\tau-\delta,\tau+\delta])})^{3/2}$$

or

$$\|f\|_{L_2^2(X\times[\tau,\tau+\delta])} \leq C(1+\|f\|_{L_2^2(X\times[\tau-\delta,\tau+\delta])})^{3/4}$$

Now

$$\|f\|_{L_2^2(X\times[\tau-\delta,\tau+\delta])} \leq C(\|f\|_{L_2^2(X\times[\tau-\delta,\tau])} + \|f\|_{L_2^2(X\times[\tau,\tau+\delta])}).$$

Then by Lemma (b) of Section 7

$$\|f\|_{L_2^2(X\times[\tau,\tau+\delta])} \leq C(1+\|f\|_{L_2^2(X\times[\tau-\delta,\tau])})^{3/4}$$

with a constant C independent of τ. There are now two cases:

<u>Case I</u>: $\omega < \infty$

Then $\|f\|_{L_2^2(X\times[\tau,\tau+\delta])}$ is a continuous function of τ for $0 \leq \tau < \omega-\delta$. Hence it is bounded for $0 \leq \tau \leq \omega - 2\delta$, and then the above result gives a bound for $\omega - 2\delta \leq \tau < \omega - \delta$ in terms of the bound for $\omega - 3\delta \leq \tau < \omega - 2\delta$.

<u>Case II</u>: $\omega = \infty$. In this case choose $\delta = 1$. Let $c_k = \|f\|_{L_2^2(X\times[k,k+1])}$. Then our result shows $c_{k+1} \leq C(1+c_k)^{3/4}$.

Then the sequence c_k is bounded by Lemma (c). If $k-1 \leq \tau \leq k$

$$\|f\|_{L_2^2(X\times[\tau,\tau+\delta])} \leq C(\|f\|_{L_2^2(X\times[k-1,k])} + \|f\|_{L_2^2(X\times[k,k+1])})$$

This completes the proof of the theorem.

<u>Corollary</u>. With a constant C independent of τ for $0 \leq \tau < \omega - \delta$ (or $0 \leq \tau < \infty$ if $\omega = \infty$)

$$\int_{\partial X\times[\tau,\tau+\delta]} |\nabla\nabla f| \leq C .$$

Proof. By the results in the proof of the last theorem

$$\int_{\partial X \times [\tau, \tau+\delta]} |\nabla\nabla f| \leq C(1 + \|f\|_{L^2_2(X \times [\tau-\delta, \tau+\delta])})^{3/2}$$

and this is bounded.

9. Let $\mathcal{C}^\infty(X/\partial)$ denote the smooth functions on X which vanish together with all their derivatives on ∂X. Let

$$\|f\|_{L^1} = \int_X |f| \; ; \; \|g\|_{L^\infty} = \sup_X |g(x)| \; ; \; \langle f,g \rangle = \int_X fg.$$

Lemma. If $f \in L^1(X)$

$$\|f\|_{L^1} = \sup\{\langle f,g \rangle : g \in \mathcal{C}^\infty(X/\partial) \text{ and } \|g\|_{L^\infty} \leq 1\}.$$

Proof. Let $\mathcal{C}^0(X)$ denote the space of continuous functions on X and $\mathcal{C}^0(X/\partial)$ the subspace of those which vanish on the boundary. Since X is a compact metric space the dual of $\mathcal{C}^0(X)$ is the space of Borel measures ν with norm $\|\nu\|$ equal to the total variation. Then $L^1(X)$ is a closed subspace and $\|f\|_{L^1}$ is the total variation of $f \in L^1(X)$ as a Borel measure. Hence

$$\|f\|_{L^1} = \sup\{\langle f,g \rangle : g \in \mathcal{C}^0(X) \text{ and } \|g\|_{L^\infty} \leq 1\}.$$

As a measure $f \in L^1(X)$ is absolutely continuous with respect to the Riemannian volume measure. Thus for any $\epsilon > 0$ we can find a neighborhood N of ∂X such that

$$\int_N |f| \leq \epsilon \|f\|_{L^1}.$$

Choose $g \in \mathcal{C}^0(X)$ with $\|g\|_{L^\infty} \leq 1$ and $\langle f,g \rangle \geq (1-\epsilon)\|f\|_{L^1}$.

Let φ be a continuous function on X with $0 \leq \varphi \leq 1$ such that $\varphi = 0$ on ∂X and $\varphi = 1$ outside of N. Then

$$\langle f, \varphi g \rangle = \langle f, g \rangle - \langle f, (1-\varphi)g \rangle .$$

But $(1-\varphi)g = 0$ outside of N and $|(1-\varphi)g| \leq 1$ so

$$|\langle f, (1-\varphi)g \rangle| \leq \int_N |f| \leq \epsilon \, \|f\|_{L^1} .$$

Thus $\langle f, \varphi g \rangle \geq (1-2\epsilon) \|f\|_{L^1}$.

Since $\epsilon > 0$ was arbitrary and $\varphi g \in \mathcal{C}^0(X/\partial)$

$$\|f\|_{L^1} = \sup\{\langle f, g \rangle : g \in \mathcal{C}^0(X/\partial) \text{ and } \|g\|_{L^\infty} \leq 1\}.$$

Now $\mathcal{C}^\infty(X/\partial)$ is dense in $\mathcal{C}^0(X/\partial)$. This completes the proof.

Let a be a smooth vector field on X and write $a\nabla g$ for $a^i \frac{\partial g}{\partial x^i}$. Let b be a smooth function on X and c a smooth function on ∂X. The following is a form of the maximum principle for Neumann boundary conditions.

<u>Theorem</u>. There exist constants B and C depending only on b and c (and independent of g, α and ω below) such that if g is continuous on $X \times [\alpha, \omega]$ and smooth for $t > \alpha$, satisfying

$$\frac{\partial g}{\partial t} = \Delta g + a\nabla g + bg \quad \text{on } X \times (\alpha, \omega]$$

$$\frac{\partial g}{\partial \nu} = cg \quad \text{on } \partial X \times (\alpha, \omega]$$

then

$$\|g\|_{L^\infty(X \times \omega)} \leq C \, e^{B(\omega - \alpha)} \, \|g\|_{L^\infty(X \times \alpha)} .$$

<u>Proof</u>. Choose a smooth function $\rho \geq 0$ on X such that $\frac{\partial \rho}{\partial \nu} = c$ on ∂X. Let $C = \max e^\rho$ and $B = \max(b + \Delta\rho - |\nabla\rho|^2 + 1)$.

Substitute $g = \tilde{g}\, e^{\rho+B(t-\alpha)}$. Then $\|\tilde{g}\|_{L^\infty(X\times\alpha)} \leq \|g\|_{L^\infty(X\times\alpha)}$ and $\|g\|_{L^\infty(X\times\omega)} \leq C\, e^{B(\omega-\alpha)} \|\tilde{g}\|_{L^\infty(X\times\omega)}$. Thus it is enough to prove

$$\|\tilde{g}\|_{L^\infty(X\times\omega)} \leq \|\tilde{g}\|_{L^\infty(X\times\alpha)}.$$

An easy calculation shows that

$$\frac{\partial\tilde{g}}{\partial t} = \Delta\tilde{g} + \tilde{a}\,\nabla\,\tilde{g} + \tilde{b}\,\tilde{g} \quad \text{on} \quad X \times (\alpha,\omega]$$

$$\frac{\partial\tilde{g}}{\partial\nu} = 0 \quad \text{on} \quad \partial X \times (\alpha,\omega]$$

where $\tilde{a} = a + 2\nabla\rho$ and $\tilde{b} = b + \Delta\rho - |\nabla\rho|^2 - B$. Thus $\tilde{b} \leq -1$. There are two possibilities, that $|\tilde{g}|$ may assume its maximum when $\tilde{g}$ is positive or when it is negative. But $-\tilde{g}$ satisfies the same equation, so it is enough to consider the case where $|\tilde{g}|$ is largest when $\tilde{g}$ is positive. Suppose the maximum of $\tilde{g}(x,t)$ is attained at a point (x^*, t^*). If $t^* = \alpha$ the proof is complete; otherwise we get a contradiction. For then we must have $\frac{\partial g}{\partial t}(x^*,t^*) \geq 0$. Moreover $\nabla g(x^*,t^*) = 0$ and $\Delta g(x^*,t^*) \leq 0$, even if $(x^*,t^*) \in \partial X \times (\alpha,\omega]$, because of the condition that the normal derivative $\frac{\partial\tilde{g}}{\partial\nu} = 0$. Unless $\tilde{g} \equiv 0$ we have $\tilde{g}(x^*,t^*) > 0$ and $\tilde{b} \leq -1$ so $\tilde{b}\,\tilde{g}\,(x^*,t^*) < 0$. This give a contradiction in the formula

$$\frac{\partial\tilde{g}}{\partial t} = \Delta\tilde{g} + \tilde{a}\,\nabla\,\tilde{g} + \tilde{b}\,\tilde{g}$$

at the point (x^*,t^*). This completes the proof.

<u>Corollary</u>. Under the same hypotheses

$$\|g\|_{L^\infty(X\times[\alpha,\omega])} \leq C\, e^{B(\omega-\alpha)}\, \|g\|_{L^\infty(X\times\alpha)}.$$

A slight modification of the proof shows the following.

<u>Theorem</u>. Let g be continuous on $X \times [\alpha,\omega]$ and smooth for $t > \alpha$. Suppose that

$$\frac{\partial g}{\partial t} \leq \Delta g + a\nabla g + bg \quad \text{on} \quad X \times (\alpha,\omega]$$

$$\frac{\partial g}{\partial \nu} \leq cg \quad \text{on} \quad \partial X \times (\alpha,\omega]$$

$$g \leq 0 \quad \text{on} \quad X \times \alpha.$$

Then $g \leq 0$ on all of $X \times [\alpha,\omega]$.

<u>Proof</u>. Again let $g = \tilde{g}\, e^{\rho+B(t-\alpha)}$. Then g is positive if and only if $\tilde{g}$ is. The preceding equalities now hold as inequalities. If $\tilde{g}$ is positive at its maximum we get a contradiction.

We are after a dual result in terms of L^1 norms.

<u>Theorem</u>. Let a be a smooth vector field and b and c smooth functions on X and ∂X. There exist constants B and C depending only on a, b and c (and independent of f, h, α and ω below) such that if f is continuous on $X \times [\alpha,\omega]$ and smooth for $t > \alpha$ and h is continuous on $\partial X \times [\alpha, \omega]$ satisfying

$$\frac{\partial f}{\partial t} = \Delta f + a\nabla f + bf \quad \text{on} \quad X \times (\alpha,\omega]$$

$$\frac{\partial f}{\partial \nu} = cf + h \quad \text{on} \quad \partial X \times (\alpha,\omega]$$

then

$$\|f\|_{L^1(X\times\omega)} \leq Ce^{B(\omega-\alpha)}\left(\|f\|_{L^1(X\times\alpha)} + \|h\|_{L^1(\partial X\times[\alpha,\omega])}\right).$$

<u>Proof</u>. Let a^* be the vector field adjoint to a, and let a_ν be the normal component of a on ∂X, so that we have a formula for integration by parts

$$\int_X a\nabla f\cdot g + \int_X f\cdot a^*\nabla g = \int_{\partial X} a_\nu fg.$$

For any choice of $g|X \times \omega$ in $\mathcal{C}^\infty(X/\partial)$ we can find a solution g of the backwards heat equation so that g is continuous on $X \times [\alpha,\omega]$ and smooth for $t < \omega$ and satisfies

$$\frac{\partial g}{\partial t} = -\Delta g + a^* \nabla g - bg \quad \text{on} \quad X \times [\alpha,\omega)$$

$$\frac{\partial g}{\partial \nu} = (c + a_\nu) g \quad \text{on} \quad \partial X \times [\alpha,\omega).$$

This follows from Section 8 of Part IV, reversing the direction of time. Now

$$\int_X \Delta f \cdot g - \int_X f \cdot \Delta g = \int_{\partial X} \frac{\partial f}{\partial \nu} \cdot g - \int_{\partial X} f \cdot \frac{\partial g}{\partial \nu}$$

by Green's theorem. Therefore for $\alpha < t < \omega$

$$\frac{d}{dt} \int_{X \times t} fg = \int_{X \times t} \frac{\partial f}{\partial t} \cdot g + f \frac{\partial g}{\partial t}$$

and, substituting the above formulas, a simple calculation yields

$$\frac{d}{dt} \int_{X \times t} fg = \int_{\partial X \times t} hg.$$

Therefore by the fundamental theorem of calculus

$$\int_{X \times \omega} fg = \int_{X \times \alpha} fg + \int_{\partial X \times [\alpha,\omega]} hg.$$

The lemma at the beginning of this section assures us that $\|f\|_{L^1(X \times \omega)}$ is the supremum of

$$\int_{X \times \omega} fg \quad \text{for } g|X \times \omega \in \mathcal{C}^\infty(X/\partial) \text{ and } \|g\|_{L^\infty(X \times \omega)} \leq 1.$$

The last theorem assures us that (with the time direction reversed) there are constants B and C depending on b and $(c + a_\nu)$ so that

$$\|g\|_{L^\infty(X\times\alpha)} \leq C\, e^{B(\omega-\alpha)} \|g\|_{L^\infty(X\times\omega)} ,$$

and applying the same estimate for $\alpha \leq \tau \leq \omega$ we can estimate

$$\|g\|_{L^\infty(\partial X\times[\alpha,\omega])} \leq C\, e^{B(\omega-\alpha)} \|g\|_{L^\infty(X\times\omega)} .$$

Therefore

$$\|f\|_{L^1(X\times\omega)} \leq C\, e^{B(\omega-\alpha)}\left(\|f\|_{L^1(X\times\alpha)} + \|h\|_{L^1(\partial X\times[\alpha,\omega])}\right).$$

Corollary. Under the same hypotheses

$$\|f\|_{L^1(X\times[\alpha,\omega])} \leq C\, e^{B(\omega-\alpha)}\left(\|f\|_{L^1(X\times\alpha)} + \|h\|_{L^1(\partial X\times[\alpha,\omega])}\right).$$

Proof. Integrate the previous estimate made for $\|f\|_{L^1(X\times\tau)}$ for $\alpha \leq \tau \leq \omega$.

Finally we prove a dual version of the Sobolev embedding theorem. Let M be a compact foliated manifold with corners, with weights $\sigma_1,\ldots,\sigma_n$ and σ their least common multiple. Recall that if $k > (\sigma/\sigma_1 + \ldots + \sigma/\sigma_n)(1/q)$ with $1 < q < \infty$ then there is a continuous inclusion $L^q_k(M) \subseteq \mathcal{C}(M)$ where $\mathcal{C}(M)$ is the continuous functions on M. This was proved in Part II.

Theorem. If $1 < p < \infty$ and $k > (\sigma/\sigma_1 + \ldots + \sigma/\sigma_n)(1-1/p)$ then

$$L^1(M) \subseteq L^p_{-k}(M)$$

with a continuous inclusion.

Proof. We have continuous inclusions

$$L^q_k(M/\partial) \to L^q_k(M) \to \mathcal{C}(M)$$

with $1/p + 1/q = 1$. By duality there are continuous maps

$$\mathcal{C}(M)^* \to L^p_{-k}(M/\partial) \to L^p_{-k}(M).$$

The kernel of the composition consists of those Borel measures which are concentrated on the boundary. Now $L^1(M)$ is a closed subspace of $\mathcal{C}(M)^*$ of measures absolutely continuous with respect to Lebesgue measure in each coordinate chart. Hence there is a continuous inclusion

$$L^1(M) \to L^p_{-k}(M).$$

If $k \geq (\sigma/\sigma_1 + \ldots + \sigma/\sigma_n)$ then this is valid for all $p < \infty$. In our case where $M = X \times [\alpha,\omega]$ we need $k \geq \dim X + 2$.

10. We use these results in the following way. The potential energy density χ satisfies the inequality

$$\frac{\partial \chi}{\partial t} \leq \Delta\chi + C\chi \qquad \text{on} \quad X \times [0,\omega)$$

$$\frac{\partial \chi}{\partial \nu} - 2m\chi = a\nabla\nabla f + b\nabla f \qquad \text{on} \quad \partial X \times [0,\omega)$$

where C is a lower bound for the Ricci curvature of X, m is the mean curvature of ∂X and $a\nabla\nabla f + b\nabla f$ is a second order linear operator with smooth coefficients, by the results of Sections 2 and 3. However we cannot use the inequality in an estimate using Gårding's inequality. Therefore we construct an auxiliary function ψ_τ satisfying the corresponding equality. Let ψ_τ be the solution on $X \times [\tau-\delta, \tau+\delta]$ of

$$\frac{\partial}{\partial t}\psi_\tau = \Delta\psi_\tau + C\psi_\tau \qquad \text{on} \quad X \times [\tau-\delta,\ \tau+\delta]$$

$$\frac{\partial}{\partial \nu}\psi_\tau - 2m\psi_\tau = a\nabla\nabla f + b\nabla f \qquad \text{on} \quad \partial X \times [\tau-\delta,\ \tau+\delta]$$

$$\psi_\tau = \chi \qquad \text{on} \quad X \times \{\tau-\delta\}.$$

By Part IV, Section 8 we know that the solution exists in $L^p_2(X \times [\tau-\delta,\ \tau+\delta])$ for any p with $\dim X + 2 < p < \infty$ and is unique, and smooth except at the corner $\partial X \times \{\tau-\delta\}$.

Lemma. $\chi \leq \psi_\tau$ on $X \times [\tau-\delta, \tau+\delta]$.

Proof. This follows from the maximum principle applied to $\chi - \psi_\tau$.

Theorem. If $k \geq \dim X+2$ then for all $p < \infty$

$$\|\psi_\tau\|_{L^p_{-k}(X\times[\tau-\delta,\tau+\delta])} \leq C$$

where C is a constant independent of τ.

Proof. By a theorem in the last section (the dual L^1 form of the maximum principle) we have

$$\|\psi_\tau\|_{L^1(X\times[\tau-\delta,\tau+\delta])} \leq C\left(\|\chi\|_{L^1(X\times\{\tau-\delta\})} + \|a\nabla\nabla f + b\nabla f\|_{L^1(\partial X\times[\tau-\delta,\tau+\delta])}\right).$$

Now

$$\|\chi\|_{L^1(X\times\{\tau-\delta\})} = \int_{X\times\{\tau-\delta\}} \chi$$

is the potential energy at time $\tau-\delta$, which is monotone decreasing in τ. Also we know that

$$\|\nabla\nabla f\|_{L^1(\partial X\times[\tau,\tau+\delta])} = \int_{\partial X\times[\tau,\tau+\delta]} |\nabla\nabla f| \leq C$$

with a constant C independent of τ by the last corollary in Section 8. Also

$$\begin{aligned}\|\nabla f\|_{L^1(\partial X\times[\tau,\tau+\delta])} &\leq C\|\nabla f\|_{L^2(\partial X\times[\tau,\tau+\delta])} \\ &\leq C\|\nabla f\|_{L^2_1(X\times[\tau,\tau+\delta])} \\ &\leq C\|f\|_{L^2_2(X\times[\tau,\tau+\delta])} \leq C\end{aligned}$$

with a constant independent of τ, by the results of Section 8. We add these over $[\tau-\delta,\tau]$ and $[\tau,\tau+\delta]$ to get estimates over $[\tau-\delta,\ \tau+\delta]$. Therefore

$$\|\psi_\tau\|_{L^p_{-k}(X\times[\tau-\delta,\tau+\delta])} \leq C\|\psi_\tau\|_{L^1(X\times[\tau-\delta,\tau+\delta])} \leq C.$$

This estimate implies that we have some kind of weak (distributional) control over the growth of ψ_τ and gives us something to interpolate off against in the following argument.

11. Theorem. For all $p < \infty$

$$\|f\|_{L^p_2(X\times[\tau,\tau+\delta])} \leq C$$

with a constant C independent of τ for $0 \leq \tau < \omega - \delta$ (or $0 \leq \tau < \infty$ if $\omega = \infty$).

Proof. Choose any $p > \dim X + 2$. We will show $\|f\|_{L^{p+1}_2(X\times[\tau,\tau+\delta])} \leq C$. Choose β with $1/p < \beta < 2/p$. In the following argument we make 3 estimates using Gårding's inequality, so we let $\epsilon = \delta/3$.

First, using Gårding's inequality as in Section 8 we have

$$\|f\|_{L^{p+1}_2(X\times[\tau,\tau+\delta])} \leq C(1 + \|\Gamma(f)\nabla f^2\|_{L^{p+1}_0(X\times[\tau-\epsilon,\tau+\delta])})$$

with a constant C independent of τ. Now $|\Gamma(f)| \leq C$ and $\frac{1}{2}|\nabla f|^2 = \chi \leq \psi_\tau$ on $[\tau-\delta,\tau+\delta]$. Therefore

$$\|\Gamma(f)\nabla f^2\|_{L^{p+1}_0(X\times[\tau-\epsilon,\tau+\delta])} \leq C\|\psi_\tau\|_{L^{p+1}_0(X\times[\tau-\epsilon,\tau+\delta])}$$

We can obtain L^{p+1}_0 by interpolating between $L^p_{1+\beta}$ and L^r_{-p} with $r = (p+1)(1+\beta)/\beta < \infty$. Therefore by interpolation

$$\| \psi_\tau \|_{L_0^{p+1}(X\times[\tau-\epsilon,\tau+\delta])}$$

$$\leq C \| \psi_\tau \|^{p/(p+1+\beta)}_{L^p_{1+\beta}(X\times[\tau-\epsilon,\tau+\delta])} \| \psi_\tau \|^{(1+\beta)/(p+1+\beta)}_{L^r_{-p}(X\times[\tau-\epsilon,\tau+\delta])}$$

However since $p > \dim X + 2$ we know by the result of Section 10 that for all $r < \infty$

$$\| \psi_\tau \|_{L^r_{-p}(X\times[\tau-\epsilon,\tau+\delta])} \leq C$$

(using the estimate over $[\tau-\epsilon,\tau+\delta-\epsilon]$ and $[\tau,\tau+\delta]$). Therefore we can omit this term.

Next we recall that ψ_τ satisfies the parabolic equation

$$\frac{\partial}{\partial t} \psi_\tau = \Delta\psi_\tau + C\psi_\tau \quad \text{on} \quad X\times[\tau-\delta,\tau+\delta]$$

and the complementing boundary condition

$$\frac{\partial}{\partial\nu} \psi_\tau - 2m\psi_\tau = a\nabla\nabla f + b\nabla f \quad \text{on} \quad \partial X\times[\tau-\delta,\tau+\delta].$$

Since $\beta > 1/p$, if $\psi_\tau \in L^p_{1+\beta}$ the boundary operator $\frac{\partial}{\partial\nu}\psi_\tau - 2m\psi_\tau | \partial X\times[\tau-\delta,\tau+\delta]$ is well defined. Therefore by Gårding's inequality for $\psi\tau$, we have for any $-k > -\infty$

$$\| \psi_\tau \|_{L^p_{1+\beta}(X\times[\tau-\epsilon,\tau+\delta])}$$

$$\leq C(\| a\nabla\nabla f + b\nabla f \|_{\partial L^p_\beta(\tau-2\epsilon,\tau+\delta])} + \| \psi_\tau \|_{L^p_{-k}(X\times[\tau-2\epsilon,\tau+\delta])})$$

with a constant C independent of τ since the equation is independent of time: that is, the operator $\frac{\partial}{\partial t} - \Delta - C$ and $\frac{\partial}{\partial\nu} - 2m|\partial$ are independent of translation in the t-direction.

But again we know from Section 10 that

$$\| \psi_\tau \|_{L^p_{-k}(X\times[\tau-2\epsilon,\tau+\delta])} \leq C$$

provided $k \geq \dim X + 2$. Therefore we can omit this term.

The other term

$$\| a\nabla\nabla f + b\nabla f\|_{\partial L^p_\beta(\partial X\times[\tau-2\epsilon,\tau+\delta])} \leq C\| f\|_{L^p_{\beta+2}(X\times[\tau-2\epsilon,\tau+\delta])}$$

We can estimate this by using Garding's inequality again: (recall $3\epsilon = \delta$)

$$\| f\|_{L^p_{\beta+2}(X\times[\tau-2\epsilon,\tau+\delta])} \leq C(1+\| \Gamma(f)\nabla f^2\|_{L^p_\beta(X\times[\tau-\delta,\tau+\delta])}) .$$

Recall that $\Gamma(f)\nabla f^2$ is a polynomial differential operator of type (2,1). By the theorem in Part IV, Section 6 with $n = 2$, $k = 1$, $p = p$, $r = \beta$, $q = p+1$, $s = 2$ we have

$$\| \Gamma(f)\nabla f^2\|_{L^p_\beta(X\times[\tau-\delta,\tau+\delta])} \leq C(1+\| f\|_{L^{p+1}_2(X\times[\tau-\delta,\tau+\delta])})^{(p+1)/p}.$$

The conditions of that theorem require that $\beta+1 < 2$ and $p(\beta+2) < (p+1)2$. The latter holds if $\beta < 2/p$, which we assumed at the beginning of the proof.

Assembling together all the estimates so far we have

$$\| f\|_{L^{p+1}_2(X\times[\tau,\tau+\delta])} \leq C(1+\| f\|_{L^{p+1}_2(X\times[\tau-\delta,\tau+\delta])})^{\alpha}$$

where $\alpha = \frac{p+1}{p} \cdot \frac{p}{p+1+\beta} = \frac{p+1}{p+1+\beta} < 1.$

We can now argue as in Section 8. First we note

$$\| f\|_{L^{p+1}_2(X\times[\tau-\delta,\tau+\delta])} \leq C(\| f\|_{L^{p+1}_2(X\times[\tau-\delta,\tau])} + \| f\|_{L^{p+1}_2(X\times[\tau,\tau+\delta])}).$$

Therefore by Lemma (b) in Section 7

$$\| f\|_{L^{p+1}_2(X\times[\tau,\tau+\delta])} \leq C(1+\| f\|_{L^{p+1}_2(X\times[\tau-\delta,\tau])})^{\alpha}.$$

There are now two cases.

<u>Case 1</u>. $\omega < \infty$.

Then we know $\| f\|_{L^{p+1}_2(X\times[\tau,\tau+\delta])}$ is continuous and hence bounded for $0 \leq \tau \leq \omega - 2\delta$. Then the estimate shows it is bounded for $\omega - 2\delta \leq \tau < \omega - \delta$ as well.

Case II. $\omega = \infty$. Then we have $\delta = 1$. Let $c_k = \| f \|_{L_2^{p+1}(X \times [k,k+1])}$. Then $c_{k+1} \leq C(1+c_k)^{\alpha}$. By Lemma (c) of Section 7 the sequence c_k is bounded. Also for any τ we can bound f on $[\tau,\tau+\delta]$ by a bound on $[k-1,k]$ and $[k,k+1]$ for some integer k.

Hence in either case

$$\| f \|_{L_2^{p+1}(X \times [\tau,\tau+\delta])} \leq C$$

with a constant C independent of τ.

12. Now we can bound all higher derivatives $\| f \|_{L_n^p(X \times [\tau,\tau+\delta])}$ as well. Because f may not be smooth at the corner $\partial X \times 0$ we restrict our estimates to $\delta \leq \tau < \omega - \delta$. Since $\omega > 4\delta$ we will have no problems. The argument is parallel to the proof of the regularity of f.

Theorem. For every $p < \infty$ and $n < \infty$ there is a constant C indpendent of τ such that

$$\| f \|_{L_n^p(X \times [\tau,\tau+\delta])} \leq C$$

for $\delta \leq \tau < \omega - \delta$ (or $1 \leq \tau < \infty$ if $\omega = \infty$).

Proof. In Section 11 we have proved this result for $n = 2$. We proceed by induction on real values of n in increments less than 1. In particular suppose $k < n + 1$, and $\| f \|_{L_n^q(X \times [\tau,\tau+\delta])} \leq C$ for $\delta \leq \tau < \omega - \delta$ and all $q < \infty$. By Gårdings inequality

$$\|f\|_{L^p_k(X\times[\tau,\tau+\delta])} \leq C(1+\|\Gamma(f)\nabla f^2\|_{L^p_{k-2}(X\times[\tau-\delta,\tau+\delta])})$$

By the theorem in Part IV Section 6 applied to the polynomial differential operator $\Gamma(f)\nabla f^2$ of type $(2,1)$ we have

$$\|\Gamma(f)\nabla f^2\|_{L^p_{k-2}(X\times[\tau-\delta,\tau+\delta])} \leq C(1+\|f\|_{L^q_n(X\times[\tau-\delta,\tau+\delta])})^{q/p}$$

provided that $k-1 < n$ and $kp < nq$. For any $p < \infty$ we can find $q < \infty$ satisfying the second. Hence the only restriction is that $k < n+1$.

Even if n is not an even integer we can bound $\|f\|_{L^q_n}$ on $X\times[\tau-\delta,\tau+\delta]$ by its norm on $X\times[\tau-\delta,\tau]$, $X\times[\tau-\delta/2,\tau+\delta/2]$ and $X\times[\tau,\tau+\delta]$, using a partition of unity on the t-axis. Therefore by the induction hypothesis

$$\|f\|_{L^q_n(X\times[\tau-\delta,\tau+\delta])} \leq C$$

with a constant independent of τ for $2\delta \leq \tau < \omega-\delta$ (or $2 \leq \tau < \infty$ if $\omega = \infty$). It follows that

$$\|f\|_{L^p_k(X\times[\tau,\tau+\delta])} \leq C$$

with a constant independent of τ for $2\delta \leq \tau < \omega-\delta$. The bound for $\delta \leq \tau \leq 2\delta$ follows by continuity. This completes the induction. (In case $\omega = \infty$ we have the same arguments for $2 \leq \tau < \infty$ and $1 \leq \tau \leq 2$).

<u>Corollary</u>. Every derivative $(\frac{\partial}{\partial t})^i \nabla^j f$ is uniformly bounded on $X\times[\delta,\omega)$ if $\omega < \infty$ and on $X\times[1,\infty)$ if $\omega = \infty$.

<u>Proof</u>. For every derivative we can find by the Sobolev embedding theorem some n and p with

$$\|(\frac{\partial}{\partial t})^i \nabla^j f\|_{L^\infty(X\times[\tau,\tau+\delta])} \leq \|f\|_{L^p_n(X\times[\tau,\tau+\delta])}$$

and the result follows.

Now a bound on $(\frac{\partial}{\partial t})^{i+1} \nabla^j f$ and $(\frac{\partial}{\partial t})^i \nabla^{j+1} f$ implies a

Lipschitz condition on $(\frac{\partial}{\partial t})^i \nabla^j f$. This proves the following result.

Corollary. If $\omega < \infty$ then the solution f of the non-linear heat equation

$$\frac{\partial f}{\partial t} = \Delta f \quad \text{on} \quad X \times [0,\omega)$$

$$f = h \quad \text{on} \quad \partial X \times [0,\omega)$$

$$f = f_0 \quad \text{on} \quad X \times 0$$

extends smoothly to a solution on $X \times [0,\omega]$, provided the Riemannian curvature $R_Y \leq 0$ and ∂Y is convex.

Let $f_\omega = f|X \times \omega$. Then $f_\omega = h$ on $\partial X \times \omega$. Therefore by the existence theorm in Part IV Section 11 we can find a solution $f \in L_2^p(X \times [\omega,\omega+\epsilon];Y)$ of the heat equation

$$\frac{\partial f}{\partial t} = \Delta f \quad \text{on} \quad X \times [\omega,\omega+\epsilon]$$

$$f = h \quad \text{on} \quad \partial X \times [\omega,\omega+\epsilon]$$

$$f = f\omega \quad \text{on} \quad X \times \omega$$

for some $\epsilon > 0$. Since the original $f \in L_2^p(X \times [0,\omega];Y)$, since it belongs to $L_2^p(X \times [0,\omega-\delta])$ and is smooth on $X \times [\delta,\omega]$, and the new $f \in L_2^p(X \times [\omega,\omega+\epsilon];Y)$ and they agree on $X \times \omega$, they define a solution $f \in L_2^p(X \times [0,\omega+\epsilon];Y)$ of the heat equation on $X \times [0,\omega+\epsilon]$, using the patching theorem of Part II, Section 12. But this contradicts the assumption that ω was chosen as large as possible. Hence we must have $\omega = \infty$. Therefore we established the following.

Theorem. Let X and Y be compact Riemannian manifolds with boundary. Suppose that the Riemannian curvature $R_Y \leq 0$ and ∂Y is convex. Let $h: \partial X \to Y$ be any smooth map and $f_0: X \to Y$ a smooth map with $f_0|\partial X = h$ in any given relative homotopy class. There exists a continuous map $f: X \times [0,\infty) \to Y$ smooth except at the corner $\partial X \times 0$ satisfying the non-linear heat equation.

$$\frac{\partial f}{\partial t} = \Delta f \qquad \text{on} \quad X \times [0,\infty)$$

$$f = h \qquad \text{on} \quad \partial X \times [0,\infty)$$

$$f = f_0 \qquad \text{on} \quad X \times 0$$

where the equation is satisfied in the L_2^p sense at the corner $\partial X \times 0$ and in the strong sense everywhere else. Moreover all derivatives $(\frac{\partial}{\partial t})^i \nabla^j f$ remain uniformly bounded as $t \to \infty$.

Let $f_t(x) = f(x,t)$ and regard f_t as a map of X into Y.

Corollary. For a suitable choice of a sequence $t_n \to \infty$ the maps $f_{t_n}: X \to Y$ converge in $C^\infty(X)$ to a harmonic map $f_\infty: X \to Y$ with $\Delta f_\infty = 0$ in the same relative homotopy class as f_0.

Proof. Since $\mathcal{C}^\infty(X)$ is a Montel space every bounded sequence has a convergent subsequence. Therefore by diagonalization we can choose a sequence $t_n \to \infty$ such that f_{t_n} converges to a smooth map f_∞: $X \to Y$ and all derivatives $(\frac{\partial}{\partial t})^i \Delta^j f_{t_n} \to (\frac{\partial}{\partial t})^i \Delta_j f_\infty$. We saw in Section 5 that when $\omega = \infty$ the total kinetic energy

$$K(f) = \int_X |\frac{\partial f}{\partial t}|^2 \to 0$$

as $t \to \infty$. Therefore we must have $\frac{\partial f_{t_n}}{\partial t} \to 0$. But $\Delta f_{t_n} \to \Delta f_\infty$ and $\frac{\partial f_{t_n}}{\partial t} = \Delta f_{t_n}$ so $\Delta f_\infty = 0$. Therefore f_∞ is harmonic. The space $\mathcal{M}_h(X,Y)$ of smooth maps of X into Y with boundary values h is easily seen to be locally arcwise connected. Therefore f_0 and f_∞ are relatively homotopic, since f_0 is relatively homotopic to t_n for all n, and f_{t_n} is relatively homotopic to f_∞ when n is large enough. This completes the proof of existence of a solution for the Dirichlet problem.

13. It is in fact true that $f_t(x) \to f_\infty(x)$ in $C^\infty(X)$ as $t \to \infty$, so that we do not need to select a subsequence. In case ∂X is empty this was proved by Hartman [9]. When ∂X is not empty there is a different and easier proof.

Lemma. There exists a constant C such that for all $f \in L_1^2(X)$ with $f|\partial X = 0$ we have

$$\int_X f^2 \leq C\int_X |\nabla f|^2.$$

Proof. If not we could find a sequence f_n in $L_1^2(X)$ with $f_n|\partial X = 0$ and

$$\int_X f_n^2 = 1 \quad \text{and} \quad \int_X |\nabla f_n|^2 \to 0.$$

Then $\|f_n\|_{L_1^2(X)} \leq C$ independent of n. Since the inclusion $L_1^2(X) \to L_0^2(X)$ is compact, by passing to a subsequence we can assume f_n converges in $L_0^2(X)$ to some function f. Also $\nabla f_n \to 0$ in $L_0^2(X)$ so $\nabla f = 0$ in the distributional sense. But 0 is continuous so f is continuously differentiable in the ordinary sense and $\nabla f = 0$. Thus f is constant. Since $f|\partial X = 0$ we must have $f = 0$. Here we use the hypothesis that ∂X is not empty. (We also tacitly assume X is connected; otherwise there is a separate argument for each component.) But we should also have $\int_X f^2 = 1$. This is a contradiction, which proves the lemma.

Theorem. Suppose ∂X is not empty. Let f be continuous on $X \times [0,\infty)$ and smooth for $t > 0$. Suppose that $f \geq 0$ on $X \times [0,\infty)$ and

$$\frac{\partial f}{\partial t} \leq \Delta f \qquad \text{on } X \times [0,\infty)$$

$$f = 0 \qquad \text{on } \partial X \times [0,\infty).$$

Then there exists an $\epsilon > 0$ and a constant C such that

$$\int_X f^2 \leq Ce^{-\epsilon t}.$$

<u>Proof</u>. We have

$$\frac{d}{dt}\int_X f^2 = \int_X f\,\frac{\partial f}{\partial t} \leq \int_X f\,\Delta f = -\int_X |\nabla f|^2$$

since $f = 0$ on $\partial X \times [0,\omega)$. By the previous lemma we have

$$\int_X |\nabla f|^2 \geq \epsilon\int_X f^2$$

for some $\epsilon > 0$. Thus we have

$$\frac{d}{dt}\int_X f^2 \leq -\epsilon\int_X f^2.$$

It follows that

$$\int_X f^2 \leq Ce^{-\epsilon t}.$$

We apply this theorem to the kinetic energy density $\varkappa = \frac{1}{2}\left|\frac{\partial f}{\partial t}\right|^2$. By the results of Section 4, we know

$$\frac{\partial\varkappa}{\partial t} \leq \Delta\varkappa \qquad \text{on } X \times (0,\infty)$$

$$\varkappa = 0 \qquad \text{on } \partial X \times (0,\infty).$$

Now $\varkappa$ may not be continuous at 0 but it surely is for $t \geq 1$. Therefore we have not only that $K(f) = \int_X \varkappa \to 0$ as $t \to \infty$, but that it decreases exponentially.

<u>Corollary</u>. If ∂X is not empty, there exists an $\epsilon > 0$ and a constant C such that for $t \geq 1$,

$$\int_X\left|\frac{\partial f}{\partial t}\right|^2 \leq Ce^{-\epsilon t}.$$

Since X is compact

$$\int_X\left|\frac{\partial f}{\partial t}\right| \leq C\left\{\int_X\left|\frac{\partial f}{\partial t}\right|^2\right\}^{1/2} \leq Ce^{-\epsilon t}$$

as well. Then for $t \leq t'$

$$\int_X |f_t(x) - f_{t'}(x)| \leq \int_X\int_t^{t'}\left|\frac{\partial f}{\partial t}\right| \leq \int_t^{t'} Ce^{-\epsilon u}\,du \leq Ce^{-\epsilon t}.$$

It follows that the $f_t(x)$ are Cauchy in $L^1(X)$ as $t \to \infty$, so the $f_t(x)$ converge in $L^1(X)$ as $t \to \infty$ to a function $f_\infty(x)$. The previous argument in Section 12 shows $f_\infty(x)$ is smooth. Now suppose the $f_t(x)$ do not converge to $f_\infty(x)$ in $\mathcal{C}^\infty(X)$. Then for some integer r and some $\delta > 0$ there is a sequence t_n with

$$\| f_{t_n} - f_\infty \|_{\mathcal{C}^r(X)} \geq \delta.$$

But the f_{t_n} are bounded in $\mathcal{C}^\infty(X)$ so by passing to a subsequence we may assume the f_{t_n} converge in $\mathcal{C}^\infty(X)$ to some function $f_\infty^* \neq f_\infty$. But the f_{t_n} do converge to f_∞ in $L^1(X)$. This is a contradition. Hence $f_t(x) \to f_\infty(x)$ in $\mathcal{C}^\infty(X)$ as $t \to \infty$.

14. The proofs for the Neumann problem and the mixed problem are almost exactly the same, perhaps a little easier. We have slightly different boundary conditions for the normal derivative of the potential energy density. For the Neumann problem we solve the nonlinear heat equation.

$$\frac{\partial f}{\partial t} = \Delta f \qquad \text{on} \quad X \times [0,\omega)$$

$$\frac{\partial f}{\partial \nu} = 0 \qquad \text{on} \quad \partial X \times [0,\omega)$$

$$f = f_0 \qquad \text{on} \quad X \times 0.$$

Again let $\chi = \frac{1}{2} |\nabla f|^2$ be the potential energy density, and let $F(\pi,\rho)$ be the second fundamental form on ∂X.

<u>Theorem</u>.

$$\frac{\partial \chi}{\partial \nu} = \langle F(\pi,\rho) \nabla_\pi f, \nabla_\rho f \rangle.$$

Proof. We compute in local coordinates using the conventions of Section 3.

$$\chi = \frac{1}{2} g^{ij} \frac{\partial f^\alpha}{\partial x^i} \frac{\partial f^\beta}{\partial x^j} h_{\alpha\beta}(f)$$

$$\frac{\partial \chi}{\partial x^n} = g^{ij}\left[\frac{\partial^2 f^\alpha}{\partial x^i \partial x^n} - \frac{\partial f^\alpha}{\partial x^k}\Gamma^k_{in} + \Gamma^\alpha_{\gamma\delta}\frac{\partial f^\gamma}{\partial x^i}\frac{\partial f^\delta}{\partial x^n}\right]\frac{\partial f^\beta}{\partial x^j} h_{\alpha\beta} .$$

In local coordinates $g^{ij} = \delta^{ij}$ and $\frac{\partial f^\beta}{\partial x^n} = 0$ so the terms vanish except for $1 \leq i = j \leq n-1$. Then $\frac{\partial^2 f^\alpha}{\partial x^i \partial x^n} = 0$ and $\frac{\partial f^\delta}{\partial x^n} = 0$, while $\frac{\partial f^\alpha}{\partial x^k}$ is zero unless $1 \leq k \leq n-1$. Using $p,q,r \ldots$ for indices restricted to $1 \leq p \leq n-1$ as before

$$\frac{\partial \chi}{\partial x^n} = -g^{ij}\,\Gamma^k_{in}\frac{\partial f^\alpha}{\partial x^k}\frac{\partial f^\beta}{\partial x^j} h_{\alpha\beta}.$$

In local coordinates $-g^{ij}\,\Gamma^k_{in}$ is the matrix of the second fundamental form. Therefore

$$\frac{\partial \chi}{\partial \nu} = \langle F(\pi,\rho)\, \nabla_\pi f, \nabla_\rho f \rangle.$$

For the kinetic energy density $\kappa = \frac{1}{2}\left|\frac{\partial f}{\partial t}\right|^2$ we have

Theorem. $\frac{\partial \kappa}{\partial \nu} = 0$ on $\partial X \times [0,\omega)$.

Proof. In local coordinates

$$\kappa = \frac{1}{2}\frac{\partial f^\alpha}{\partial t}\frac{\partial f^\beta}{\partial t} h_{\alpha\beta}$$

$$\frac{\partial \kappa}{\partial x^n} = \left[\frac{\partial f^\alpha}{\partial x^n \partial t} - \Gamma^\alpha_{\gamma\delta}\frac{\partial f^\gamma}{\partial x^n}\frac{\partial f^\delta}{\partial t}\right]\frac{\partial f^\beta}{\partial t} h_{\alpha\beta} .$$

But $\frac{\partial f^\gamma}{\partial x^n}$ and $\frac{\partial^2 f^\alpha}{\partial x^n \partial t}$ are zero on $\partial X \times [0,\omega)$.

Now the proof proceeds in the same way, except that instead of the boundary condition $\frac{\partial \chi}{\partial \nu} - 2m\chi = a\nabla\nabla f + bf$ we have instead a boundary condition $\frac{\partial \chi}{\partial \nu} = a\nabla f^2$. But in Section 6 we have already seen the estimate

$$\int_{\partial X \times T} |\nabla f|^2 \leq C\| f\|^{4/3}_{L_2^2(X\times T)}$$

used to handle such a term. Therefore there are no new problems in the proof. Another small change is that in showing the solution is unique and remains inside Y as in Part IV, Sections 3 and 4 we must replace the Dirichlet condition $\rho = 0$ on $\partial X \times [0,\omega)$ with the Neumann condition $\frac{\partial \rho}{\partial \nu} = 0$ on $\partial X \times [0,\omega)$. The same holds in all estimates and existence theorems for f itself.

15. Finally consider the mixed problem

$$\Delta f = 0 \qquad \text{on } X$$

$$f(\partial X) \subseteq \partial Y$$

$$\frac{\partial f}{\partial \nu} \perp \partial Y \qquad \text{on } \partial X.$$

Here we must assume ∂Y is totally geodesic in order to get an estimate on the normal derivative of the potential energy density. To make the problem more tractable we embed the double $\tilde{Y}$ of Y in the Euclidean space R^N in such a way that the last coordinate function y^N is the distance to ∂Y, at least in a neighborhood of ∂Y. We then extend the metric on Y to $\tilde{Y}$ and then all of R^N as before. We solve for a map $f\colon X \times [0,\omega) \to R^N$ satisfying

$$\frac{\partial f}{\partial t} = \Delta f \quad \text{on} \quad X \times [0,\omega)$$

$$\left.\begin{array}{l} f^N = 0 \\ \dfrac{\partial f^\alpha}{\partial \nu} = 0 \quad \text{for} \quad 1 \leq \alpha \leq N-1 \end{array}\right\} \text{on} \quad \partial X \times [0,\omega)$$

$$f = f_0 \quad \text{on} \quad X \times 0.$$

In proving uniqueness (which by the symmetry of the metric on a tubular neighborhood shows as before that the solution remains in $\tilde{Y}$) we need to estimate the normal derivative $\frac{\partial \rho}{\partial \nu}$ where $\rho = \sigma(f_1,f_2)$ and $\sigma(y_1,y_2) = \frac{1}{2}d(y_1,y_2)^2$. We have

$$\frac{\partial \rho}{\partial \nu} = \frac{\partial \sigma}{\partial y_1^\alpha}\frac{\partial f_1^\alpha}{\partial \nu} + \frac{\partial \sigma}{\partial y_2^\alpha}\frac{\partial f_2^\alpha}{\partial \nu}.$$

The terms with $\alpha < N$ are all zero on $\partial X \times [0,\omega)$. When $\alpha = N$ we use the following.

<u>Lemma</u>. If $y_1^N = y_2^N = 0$ then

$$\left|\frac{\partial \sigma}{\partial y_1^N}(y_1,y_2)\right| \leq C\,\sigma(y_1,y_2).$$

<u>Proof</u>. Since $\sigma = \frac{1}{2}d(y_1,y_2)^2$ we have

$$\frac{\partial \sigma}{\partial y_1^N} = d(y_1,y_2)\cdot\frac{\partial d(y_1,y_2)}{\partial y_1^N}.$$

Let γ be the unit tangent vector at y_1 to the geodesic from y_1 to y_2, and let ν be the unit normal vector. Then

$$\frac{\partial d(y_1,y_2)}{\partial y_1^N} = \langle \gamma, \nu \rangle.$$

But y_1 and y_2 both lie on ∂Y and ν is normal so $|\langle \gamma,\nu\rangle| \leq C\,d(y_1,y_2)$. Then

$$\left|\frac{\partial \sigma}{\partial y_1^N}(y_1,y_2)\right| \leq C d(y_1,y_2)^2 \leq C\sigma(y_1,y_2).$$

We have the same for $\frac{\partial \sigma}{\partial y_2^N}$, and $\frac{\partial f_1^N}{\partial \nu}$ and $\frac{\partial f_2^N}{\partial \nu}$ are continuous and hence bounded. This shows that we have an estimate $\frac{\partial\rho}{\partial\nu} \leq C\rho$ on $\partial X \times [0,\omega)$ which is enough to prove uniqueness.

Let $\chi = \frac{1}{2}|\nabla f|^2$ be the kinetic energy density. Then

$\chi = \frac{1}{2} g^{ij} \frac{\partial f^\alpha}{\partial x^i} \frac{\partial f^\beta}{\partial x^j} h_{\alpha\beta}(f)$ and

$$\frac{\partial \chi}{\partial x^n} = g^{ij}\left[\frac{\partial^2 f^\alpha}{\partial x^i \partial x^n} - \frac{\partial f^\alpha}{\partial x^k} \Gamma^k_{in} + \Gamma^\alpha_{\gamma\delta} \frac{\partial f^\gamma}{\partial x^i} \frac{\partial f^\delta}{\partial x^n} \right] \frac{\partial f^\beta}{\partial x^j} h_{\alpha\beta} .$$

We choose coordinates $\{x^1,\ldots,x^n\}$ so that $\partial X = \{x^n = 0\}$ and $\{y^1,\ldots,y^N\}$ so that $\partial Y = \{y^N = 0\}$. Then $f^N = 0$ and $\frac{\partial f^\alpha}{\partial x^n} = 0$ for $\alpha < n$ on the boundary. Choose reference points $x \in \partial X$ and $y = f(x) \in \partial Y$. The condition that ∂Y is totally geodesic is that we can choose coordinates as above and simultaneously have all $\Gamma^\alpha_{\beta\gamma} = 0$ at the point $y \in \partial Y$. We can also make $g^{ij} = \delta^{ij}$ at x and $h_{\alpha\beta} = \delta_{\alpha\beta}$ at y. Then the terms with $\alpha = \beta < N$ produce a contribution

$$\langle F(\pi,\rho)\nabla_\pi f,\ \nabla_\rho f\rangle$$

as in the Neumann problem; while the terms $\alpha = \beta = N$, using the equation for f^N since $\frac{\partial f^N}{\partial t}$ is zero on the boundary, produce a contribution

$$m|\nabla_\nu f|^2$$

as in the Dirichlet problem. Therefore we have an equation

$$\frac{\partial \chi}{\partial \nu} = m|\nabla_\nu f|^2 + \langle F(\pi,\rho)\nabla_\pi f,\ \nabla_\rho f\rangle.$$

Finally if $\varkappa = \frac{1}{2}\left|\frac{\partial f}{\partial t}\right|^2$ is the kinetic energy $\varkappa = \frac{1}{2} \frac{\partial f^\alpha}{\partial t} \frac{\partial f^\beta}{\partial t} h_{\alpha\beta}$

$$\frac{\partial \varkappa}{\partial x^n} = \left[\frac{\partial^2 f^\alpha}{\partial x^n \partial t} - \Gamma^\alpha_{\gamma\delta} \frac{\partial f^\gamma}{\partial x^n} \frac{\partial f^\delta}{\partial t} \right] \frac{\partial f^\beta}{\partial t} h_{\alpha\beta} .$$

Again we get zero in case $\alpha = \beta < N$ or $\alpha = \beta = N$. Thus $\frac{\partial \varkappa}{\partial \nu} = 0$ on $\partial X \times [0,\infty)$. The rest of the proof proceeds the same.

Bibliography

[1] Agmon, S., Douglis, A., and Nirenberg, L., Estimates near the boundary for solutions of elliptic partial differential equations satisfying general boundary conditions, I, Comm. Pure Appl. Math., Vol. 12, 1959, pp. 623-727, and II, ibid, Vol. 17, 1964, pp. 35-92.

[2] Calderón, A.P., Intermediate spaces and interpolation, the complex method, Studia Math., Vol. 24, 1964, pp. 113-190.

[3] Eells, J., A setting for global analysis, Bull. Amer. Math. Soc., Vol. 72, 1966, pp. 739-807.

[4] Eells, J., and Sampson, J.H., Harmonic mappings of Riemannian manifolds, Amer. J. Math., Vol. 86, 1964, pp. 109-160.

[5] Eliasson, H.I., On the geometry of manifolds of maps, J. Diff. Geom., Vol. 1, 1967, pp. 169-194.

[6] Eliasson, H.I., Variational integrals in fibre bundles, A.M.S. Proc. Symposia in Pure Math., Vol. 16, 1968, pp. 67-89.

[7] Fabes, E.B., and Riviere, N.M., Singular integrals with mixed homogeneity, Studia Math., Vol. 27, 1966, pp. 19-38.

[8] Friedman, A., Partial differential equations of parabolic type, Prentice-Hall Inc., Englewood Cliffs, N.J., 1964, pp. 347.

[9] Hartman, P., On homotopic harmonic maps, Can. J. Math., Vol. 19, 1967, pp. 673-687.

[10] Hörmander, L., Linear partial differential operators, Springer-Verlag, Berlin, 1964, pp. 284.

[11] Jones, B.L., Lipschitz spaces and the heat equation, J. Math. Mech., Vol. 18, 1968, pp. 379-410.

[12] Jones, B.F., and Tu, C.C., Embedding and continuation theorems for parabolic function spaces, Amer. J. Math., Vol. 92, 1970, pp. 857-868.

[13] Lang, S., Introduction to differentiable manifolds, Interscience, New York, 1962, pp. 126.

[14] Lichnerowitz, A., Applications Harmoniques et Varietes Kähleriennes, Symposia Mathematica III, pp. 341-402.

[15] Palais, R.S., Foundations of global non-linear analysis, Benjamin, New York, 1968.

[16] Petree, J., Espaces d'interpolation et theòreme de Soboleff, Ann. Inst. Fourier 16, 1966, pp. 279-317.

[17] Petree, J., Sur les espaces de Besov, C.R. Acad. Sci., Paris 264, 1967, A281-A283.

[18] Petree, J., A theory of interpolation of normed spaces, Notas de Matemática No. 39, Instituto de Matemática Pura e Aplicada, Rio de Janeior, 1968, pp. 83.

[19] Rivière, N.M., Singular integrals and multiplier operators, Arkiv für Mat., Vol. 9, 1971, pp. 243-278.

[20] Rudin, W., Real and complex analysis, McGraw-Hill, New York, 1966, pp. 412.

[21] Smith, R.L., Harmonic mappings of spheres, thesis, University of Warwick, 1972, pp. 127.

[22] Stein, E. M., Singular integrals and differentiability properties of functions, Princeton U. Press, 1970.

[23] Strichartz, R.S., Multipliers on fractional Sobolev spaces, J. Math. Mech., Vol. 16, 1967, pp. 1031-1060.

[24] Taibleson, M.H., The preservation of Lipschitz spaces under singular integral operators, Studia Math., Vol. 24, 1963, pp. 105-111.

[25] Taibleson, M.H., On the theory of Lipschitz spaces of distributions on Euclidean n-space, I, J. Math. Mech., Vol. 13, 1964, pp. 407-480; II, ibid, Vol. 14, 1965, pp. 821-840; III, ibid, Vol. 15, 1966, pp. 973-981.

[26] Torchinsky, A., Singular integrals in the spaces $\Lambda(B,X)$, Studia Math., to appear.

[27] Uhlenbeck, K., Regularity theorems for solutions of elliptic polynomial equations, A.M.S. Proc. Symposia in Pure Math., Vol. 16, 1968, pp. 225-232.

[28] Uhlenbeck, K., Morse theory on Banach manifolds, Bull. A.M.S., Vol. 76, 1970, pp. 105-106.

[29] Uhlenbeck, K., Harmonic maps: a direct method in the calculus of variations, Bull. A.M.S., Vol. 76, 1970, pp. 1082-1087.

Vol. 309: D. H. Sattinger, Topics in Stability and Bifurcation Theory. VI, 190 pages. 1973 DM 20,-

Vol 310: B Iversen, Generic Local Structure of the Morphisms in Commutative Algebra. IV, 108 pages. 1973. DM 18,-

Vol. 311: Conference on Commutative Algebra. Edited by J. W. Brewer and E. A. Rutter. VII, 251 pages. 1973. DM 24,-

Vol. 312: Symposium on Ordinary Differential Equations. Edited by W. A Harris, Jr. and Y. Sibuya. VIII, 204 pages. 1973. DM 22,-

Vol. 313: K. Jörgens and J. Weidmann, Spectral Properties of Hamiltonian Operators. III, 140 pages. 1973. DM 18,-

Vol. 314: M. Deuring, Lectures on the Theory of Algebraic Functions of One Variable. VI, 151 pages. 1973 DM 18,-

Vol. 315: K. Bichteler, Integration Theory (with Special Attention to Vector Measures). VI, 357 pages. 1973. DM 29,-

Vol. 316: Symposium on Non-Well-Posed Problems and Logarithmic Convexity. Edited by R J Knops V, 176 pages 1973 DM 20,-

Vol 317: Séminaire Bourbaki - vol. 1971/72 Exposés 400-417. IV, 361 pages. 1973. DM 29,-

Vol. 318: Recent Advances in Topological Dynamics. Edited by A. Beck. VIII, 285 pages. 1973. DM 27,-

Vol. 319: Conference on Group Theory. Edited by R W. Gatterdam and K. W. Weston. V, 188 pages. 1973. DM 20,-

Vol. 320: Modular Functions of One Variable I. Edited by W. Kuyk. V, 195 pages. 1973 DM 20,-

Vol. 321: Séminaire de Probabilités VII. Edité par P. A. Meyer. VI, 322 pages. 1973. DM 29,-

Vol. 322: Nonlinear Problems in the Physical Sciences and Biology. Edited by I Stakgold, D D Joseph and D H Sattinger VIII, 357 pages 1973 DM 29,-

Vol. 323: J. L. Lions, Perturbations Singulières dans les Problèmes aux Limites et en Contrôle Optimal. XII, 645 pages. 1973 DM 46,-

Vol. 324: K. Kreith, Oscillation Theory. VI, 109 pages. 1973 DM 18,-

Vol. 325: C.-C. Chou, La Transformation de Fourier Complexe et L'Equation de Convolution IX, 137 pages 1973 DM 18,-

Vol. 326: A Robert, Elliptic Curves. VIII, 264 pages 1973. DM 24,-

Vol. 327: E. Matlis, One-Dimensional Cohen-Macaulay Rings XII, 157 pages. 1973. DM 20,-

Vol 328: J. R. Büchi and D. Siefkes, The Monadic Second Order Theory of All Countable Ordinals. VI, 217 pages. 1973. DM 22,-

Vol. 329: W. Trebels, Multipliers for (C, α)-Bounded Fourier Expansions in Banach Spaces and Approximation Theory. VII, 103 pages. 1973. DM 18,-

Vol 330: Proceedings of the Second Japan-USSR Symposium on Probability Theory. Edited by G. Maruyama and Yu. V. Prokhorov. VI, 550 pages. 1973. DM 40,-

Vol. 331: Summer School on Topological Vector Spaces. Edited by L. Waelbroeck. VI, 226 pages. 1973. DM 22,-

Vol 332: Séminaire Pierre Lelong (Analyse) Année 1971-1972. V, 131 pages. 1973 DM 18,-

Vol. 333: Numerische, insbesondere approximationstheoretische Behandlung von Funktionalgleichungen. Herausgegeben von R. Ansorge und W. Törnig. VI, 296 Seiten. 1973. DM 27,-

Vol. 334: F. Schweiger, The Metrical Theory of Jacobi-Perron Algorithm. V, 111 pages. 1973 DM 18,-

Vol. 335: H. Huck, R. Roitzsch, U. Simon, W. Vortisch, R. Walden, B. Wegner und W. Wendland, Beweismethoden der Differentialgeometrie im Großen. IX, 159 Seiten. 1973. DM 20,-

Vol. 336: L'Analyse Harmonique dans le Domaine Complexe. Edité par E. J. Akutowicz. VIII, 169 pages. 1973. DM 20,-

Vol. 337: Cambridge Summer School in Mathematical Logic. Edited by A. R. D. Mathias and H. Rogers. IX, 660 pages. 1973. DM 46,-

Vol. 338: J. Lindenstrauss and L. Tzafriri, Classical Banach Spaces IX, 243 pages 1973. DM 24,-

Vol. 339: G. Kempf, F. Knudsen, D. Mumford and B. Saint-Donat, Toroidal Embeddings I. VIII, 209 pages. 1973. DM 22,-

Vol. 340: Groupes de Monodromie en Géométrie Algébrique. (SGA 7 II). Par P. Deligne et N. Katz. X, 438 pages. 1973. DM 44,-

Vol. 341: Algebraic K-Theory I, Higher K-Theories. Edited by H. Bass. XV, 335 pages. 1973. DM 29,-

Vol. 342: Algebraic K-Theory II, "Classical" Algebraic K-Theory, and Connections with Arithmetic. Edited by H. Bass. XV, 527 pages. 1973. DM 40,-

Vol. 343: Algebraic K-Theory III, Hermitian K-Theory and Geometric Applications. Edited by H. Bass. XV, 572 pages. 1973. DM 40,-

Vol. 344: A. S. Troelstra (Editor), Metamathematical Investigation of Intuitionistic Arithmetic and Analysis. XVII, 485 pages. 1973 DM 38,-

Vol. 345: Proceedings of a Conference on Operator Theory. Edited by P A. Fillmore. VI, 228 pages 1973. DM 22,-

Vol. 346: Fučík et al., Spectral Analysis of Nonlinear Operators. II, 287 pages 1973. DM 26,-

Vol. 347: J. M. Boardman and R. M. Vogt, Homotopy Invariant Algebraic Structures on Topological Spaces. X, 257 pages. 1973. DM 24,-

Vol. 348: A. M. Mathai and R. K. Saxena, Generalized Hypergeometric Functions with Applications in Statistics and Physical Sciences. VII, 314 pages. 1973. DM 26,-

Vol. 349: Modular Functions of One Variable II Edited by W. Kuyk and P. Deligne. V, 598 pages. 1973. DM 38,-

Vol. 350: Modular Functions of One Variable III. Edited by W. Kuyk and J.-P. Serre. V, 350 pages. 1973. DM 26,-

Vol. 351: H. Tachikawa, Quasi-Frobenius Rings and Generalizations. XI, 172 pages. 1973. DM 20,-

Vol. 352: J. D. Fay, Theta Functions on Riemann Surfaces. V, 137 pages. 1973. DM 18,-

Vol. 353: Proceedings of the Conference on Orders, Group Rings and Related Topics. Organized by J. S. Hsia, M. L. Madan and T. G. Ralley. X, 224 pages. 1973. DM 22,-

Vol. 354: K J. Devlin, Aspects of Constructibility. XII, 240 pages. 1973. DM 24,-

Vol. 355: M. Sion, A Theory of Semigroup Valued Measures V, 140 pages. 1973. DM 18,-

Vol. 356: W. L. J. van der Kallen, Infinitesimally Central-Extensions of Chevalley Groups. VII, 147 pages. 1973. DM 18,-

Vol. 357: W. Borho, P. Gabriel und R. Rentschler, Primideale in Einhüllenden auflösbarer Lie-Algebren. V, 182 Seiten. 1973. DM 20,-

Vol. 358: F. L. Williams, Tensor Products of Principal Series Representations. VI, 132 pages. 1973. DM 18,-

Vol. 359: U Stammbach, Homology in Group Theory. VIII, 183 pages. 1973 DM 20,-

Vol. 360: W. J. Padgett and R. L. Taylor, Laws of Large Numbers for Normed Linear Spaces and Certain Fréchet Spaces. VI, 111 pages. 1973. DM 18,-

Vol. 361: J. W. Schutz, Foundations of Special Relativity: Kinematic Axioms for Minkowski Space Time. XX, 314 pages. 1973. DM 26,-

Vol. 362: Proceedings of the Conference on Numerical Solution of Ordinary Differential Equations. Edited by D. Bettis. VIII, 490 pages 1974. DM 34,-

Vol. 363: Conference on the Numerical Solution of Differential Equations. Edited by G. A. Watson. IX, 221 pages. 1974 DM 20,-

Vol. 364: Proceedings on Infinite Dimensional Holomorphy. Edited by T. L. Hayden and T. J. Suffridge. VII, 212 pages 1974. DM 20,-

Vol. 365: R. P. Gilbert, Constructive Methods for Elliptic Equations. VII, 397 pages. 1974. DM 26,-

Vol. 366: R. Steinberg, Conjugacy Classes in Algebraic Groups (Notes by V V. Deodhar). VI, 159 pages. 1974. DM 18,-

Vol. 367: K Langmann und W. Lütkebohmert, Cousinverteilungen und Fortsetzungssätze. VI, 151 Seiten. 1974. DM 16,-

Vol. 368: R. J. Milgram, Unstable Homotopy from the Stable Point of View. V, 109 pages. 1974. DM 16,-

Vol. 369: Victoria Symposium on Nonstandard Analysis. Edited by A. Hurd and P. Loeb. XVIII, 339 pages. 1974. DM 26,-

Vol. 370: B. Mazur and W. Messing, Universal Extensions and One Dimensional Crystalline Cohomology. VII, 134 pages. 1974. DM 16,–

Vol. 371: V. Poenaru, Analyse Différentielle. V, 228 pages. 1974. DM 20,–

Vol. 372: Proceedings of the Second International Conference on the Theory of Groups 1973. Edited by M. F. Newman. VII, 740 pages. 1974. DM 48,–

Vol. 373: A. E. R. Woodcock and T. Poston, A Geometrical Study of the Elementary Catastrophes. V, 257 pages. 1974. DM 22,–

Vol. 374: S. Yamamuro, Differential Calculus in Topological Linear Spaces. IV, 179 pages. 1974. DM 18,–

Vol. 375: Topology Conference 1973. Edited by R. F. Dickman Jr. and P. Fletcher. X, 283 pages. 1974. DM 24,–

Vol. 376: D. B. Osteyee and I. J. Good, Information, Weight of Evidence, the Singularity between Probability Measures and Signal Detection. XI, 156 pages. 1974. DM 16,–

Vol. 377: A. M. Fink, Almost Periodic Differential Equations. VIII, 336 pages. 1974. DM 26,–

Vol. 378: TOPO 72 – General Topology and its Applications. Proceedings 1972. Edited by R. Alò, R. W. Heath and J. Nagata. XIV, 651 pages. 1974. DM 50,–

Vol. 379: A. Badrikian et S. Chevet, Mesures Cylindriques, Espaces de Wiener et Fonctions Aléatoires Gaussiennes. X, 383 pages. 1974. DM 32,–

Vol. 380: M. Petrich, Rings and Semigroups. VIII, 182 pages. 1974. DM 18,–

Vol. 381: Séminaire de Probabilités VIII. Edité par P. A. Meyer. IX, 354 pages. 1974. DM 32,–

Vol. 382: J. H. van Lint, Combinatorial Theory Seminar Eindhoven University of Technology. VI, 131 pages. 1974. DM 18,–

Vol. 383: Séminaire Bourbaki – vol. 1972/73. Exposés 418-435. IV, 334 pages. 1974. DM 30,–

Vol. 384: Functional Analysis and Applications, Proceedings 1972. Edited by L. Nachbin. V, 270 pages. 1974. DM 22,–

Vol. 385: J. Douglas Jr. and T. Dupont, Collocation Methods for Parabolic Equations in a Single Space Variable (Based on C^1-Piecewise-Polynomial Spaces). V, 147 pages. 1974. DM 16,–

Vol. 386: J. Tits, Buildings of Spherical Type and Finite BN-Pairs. IX, 299 pages. 1974. DM 24,–

Vol. 387: C. P. Bruter, Eléments de la Théorie des Matroïdes. V, 138 pages. 1974. DM 18,–

Vol. 388: R. L. Lipsman, Group Representations. X, 166 pages. 1974. DM 20,–

Vol. 389: M.-A. Knus et M. Ojanguren, Théorie de la Descente et Algèbres d' Azumaya. IV, 163 pages. 1974. DM 20,–

Vol. 390: P. A. Meyer, P. Priouret et F. Spitzer, Ecole d'Eté de Probabilités de Saint-Flour III – 1973. Edité par A. Badrikian et P.-L. Hennequin. VIII, 189 pages. 1974. DM 20,–

Vol. 391: J. Gray, Formal Category Theory: Adjointness for 2-Categories. XII, 282 pages. 1974. DM 24,–

Vol. 392: Géométrie Différentielle, Colloque, Santiago de Compostela, Espagne 1972. Edité par E. Vidal. VI, 225 pages. 1974. DM 20,–

Vol. 393: G. Wassermann, Stability of Unfoldings. IX, 164 pages. 1974. DM 20,–

Vol. 394: W. M. Patterson 3rd, Iterative Methods for the Solution of a Linear Operator Equation in Hilbert Space – A Survey. III, 183 pages. 1974. DM 20,–

Vol. 395: Numerische Behandlung nichtlinearer Integrodifferential- und Differentialgleichungen. Tagung 1973. Herausgegeben von R. Ansorge und W. Törnig. VII, 313 Seiten. 1974. DM 28,–

Vol. 396: K. H. Hofmann, M. Mislove and A. Stralka, The Pontryagin Duality of Compact O-Dimensional Semilattices and its Applications. XVI, 122 pages. 1974. DM 18,–

Vol. 397: T. Yamada, The Schur Subgroup of the Brauer Group. V, 159 pages. 1974. DM 18,–

Vol. 398: Théories de l'Information, Actes des Rencontres de Marseille-Luminy, 1973. Edité par J. Kampé de Fériet et C. Picard. XII, 201 pages. 1974. DM 23,–

Vol. 399: Functional Analysis and its Applications, Proceedings 1973. Edited by H. G. Garnir, K. R. Unni and J. H. Williamson. XVII, 569 pages. 1974. DM 44,–

Vol. 400: A Crash Course on Kleinian Groups – San Francisco 1974. Edited by L. Bers and I. Kra. VII, 130 pages. 1974. DM 18,–

Vol. 401: F. Atiyah, Elliptic Operators and Compact Groups. V, 93 pages. 1974. DM 18,–

Vol. 402: M. Waldschmidt, Nombres Transcendants. VIII, 277 pages. 1974. DM 25,–

Vol. 403: Combinatorial Mathematics – Proceedings 1972. Edited by D. A. Holton. VIII, 148 pages. 1974. DM 18,–

Vol. 404: Théorie du Potentiel et Analyse Harmonique. Edité par J. Faraut. V, 245 pages. 1974. DM 25,–

Vol. 405: K. Devlin and H. Johnsbråten, The Souslin Problem. VIII, 132 pages. 1974. DM 18,–

Vol. 406: Graphs and Combinatorics – Proceedings 1973. Edited by R. A. Bari and F. Harary. VIII, 355 pages. 1974. DM 30,–

Vol. 407: P. Berthelot, Cohomologie Cristalline des Schémas de Caracteristique p > o. VIII, 598 pages. 1974. DM 44,–

Vol. 408: J. Wermer, Potential Theory. VIII, 146 pages. 1974. DM 18,–

Vol. 409: Fonctions de Plusieurs Variables Complexes, Séminaire François Norguet 1970–1973. XIII, 612 pages. 1974. DM 47,–

Vol. 410: Séminaire Pierre Lelong (Analyse) Année 1972–1973. VI, 181 pages. 1974. DM 18,–

Vol. 411: Hypergraph Seminar. Ohio State University, 1972. Edited by C. Berge and D. Ray-Chaudhuri. IX, 287 pages. 1974. DM 28,–

Vol. 412: Classification of Algebraic Varieties and Compact Complex Manifolds. Proceedings 1974. Edited by H. Popp. V, 333 pages. 1974. DM 30,–

Vol. 413: M. Bruneau, Variation Totale d'une Fonction. XIV, 332 pages. 1974. DM 30,–

Vol. 414: T. Kambayashi, M. Miyanishi and M. Takeuchi, Unipotent Algebraic Groups. VI, 165 pages. 1974. DM 20,–

Vol. 415: Ordinary and Partial Differential Equations, Proceedings of the Conference held at Dundee, 1974. XVII, 447 pages. 1974. DM 37,–

Vol. 416: M. E. Taylor, Pseudo Differential Operators. IV, 155 pages. 1974. DM 18,–

Vol. 417: H. H. Keller, Differential Calculus in Locally Convex Spaces. XVI, 131 pages. 1974. DM 18,–

Vol. 418: Localization in Group Theory and Homotopy Theory and Related Topics. Battelle Seattle 1974 Seminar. Edited by P. J. Hilton. VI, 171 pages. 1974. DM 20,–

Vol. 419: Topics in Analysis – Proceedings 1970. Edited by O. E. Lehto, I. S. Louhivaara, and R. H. Nevanlinna. XIII, 391 pages. 1974. DM 35,–

Vol. 420: Category Seminar. Proceedings, Sydney Category Theory Seminar 1972/73. Edited by G. M. Kelly. VI, 375 pages. 1974. DM 32,–

Vol. 421: V. Poénaru, Groupes Discrets. VI, 216 pages. 1974. DM 23,–

Vol. 422: J.-M. Lemaire, Algèbres Connexes et Homologie des Espaces de Lacets. XIV, 133 pages. 1974. DM 23,–

Vol. 423: S. S. Abhyankar and A. M. Sathaye, Geometric Theory of Algebraic Space Curves. XIV, 302 pages. 1974. DM 28,–

Vol. 424: L. Weiss and J. Wolfowitz, Maximum Probability Estimators and Related Topics. V, 106 pages. 1974. DM 18,–

Vol. 425: P. R. Chernoff and J. E. Marsden, Properties of Infinite Dimensional Hamiltonian Systems. IV, 160 pages. 1974. DM 20,–

Vol. 426: M. L. Silverstein, Symmetric Markov Processes. IX, 287 pages. 1974. DM 28,–

Vol. 427: H. Omori, Infinite Dimensional Lie Transformation Groups. XII, 149 pages. 1974. DM 18,–

Vol. 428: Algebraic and Geometrical Methods in Topology. Proceedings 1973. Edited by L. F. McAuley. XI, 280 pages. 1974. DM 28,–

Vol. 429: L. Cohn, Analytic Theory of the Harish-Chandra C-Function. IV, 154 pages. 1974. DM 18,–

Vol. 430: Constructive and Computational Methods for Differential and Integral Equations. Symposium 1974. Edited by D. L. Colton and R. P. Gilbert. VII, 476 pages. 1974. DM 37,–

Vol. 431: Séminaire Bourbaki – vol. 1973/74. Exposés 436–452. IV, 347 pages. 1975. DM 30,–

Vol. 432: R. P. Pflug, Holomorphiegebiete, pseudokonvexe Gebiete und das Levi-Problem. VI, 210 Seiten. 1975. DM 23,–

Vol. 433: W. G. Faris, Self-Adjoint Operators. VII, 115 pages. 1975. DM 18,–

Vol. 434: P. Brenner, V. Thomée, and L. B. Wahlbin, Besov Spaces and Applications to Difference Methods for Initial Value Problems. II, 154 pages. 1975. DM 18,–

Vol. 435: C. F. Dunkl and D. E. Ramirez, Representations of Commutative Semitopological Semigroups. VI, 181 pages. 1975. DM 20,–

Vol. 436: L. Auslander and R. Tolimieri, Abelian Harmonic Analysis, Theta Functions and Function Algebras on a Nilmanifold. V, 99 pages. 1975. DM 18,–

Vol. 437: D. W. Masser, Elliptic Functions and Transcendence. XIV, 143 pages. 1975. DM 18,–

Vol. 438: Geometric Topology, Proceedings 1974. Edited by L. C. Glaser and T. B. Rushing. X, 459 pages. 1975. DM 37,–

Vol. 439: K. Ueno, Classification Theory of Algebraic Varieties and Compact Complex Spaces. XIX, 278 pages. 1975 DM 28,–

Vol. 440: R. K. Getoor, Markov Processes: Ray Processes and Right Processes. V, 118 pages. 1975. DM 18,–

Vol. 441: N. Jacobson, PI-Algebras. An Introduction. V, 115 pages. 1975. DM 18,–

Vol. 442: C. H. Wilcox, Scattering Theory for the d'Alembert Equation in Exterior Domains. III, 184 pages. 1975. DM 20,–

Vol. 443: M. Lazard, Commutative Formal Groups. II, 236 pages. 1975. DM 23,–

Vol. 444: F. van Oystaeyen, Prime Spectra in Non-Commutative Algebra. V, 128 pages. 1975. DM 18,–

Vol. 445: Model Theory and Topoi. Edited by F. W. Lawvere, C. Maurer and G. C. Wraith. III, 354 pages. 1975. DM 30,–

Vol. 446: Partial Differential Equations and Related Topics, Proceedings 1974. Edited by J. A. Goldstein. IV, 389 pages. 1975. DM 32,–

Vol. 447: S. Toledo, Tableau Systems for First Order Number Theory and Certain Higher Order Theories. III, 339 pages. 1975. DM 30,–

Vol. 448: Spectral Theory and Differential Equations. Proceedings 1974. Edited by W. N. Everitt. XII, 321 pages. 1975. DM 30,–

Vol. 449: Hyperfunctions and Theoretical Physics. Proceedings 1973. Edited by F. Pham. IV, 218 pages. 1975. DM 23,–

Vol. 450: Algebra and Logic. Proceedings 1974. Edited by J. N. Crossley. VIII, 307 pages. 1975. DM 28,–

Vol. 451: Probabilistic Methods in Differential Equations. Proceedings 1974. Edited by M. A. Pinsky. VII, 162 pages. 1975. DM 20,–

Vol. 452: Combinatorial Mathematics III. Proceedings 1974. Edited by Anne Penfold Street and W. D. Wallis. IX, 233 pages. 1975. DM 23,–

Vol. 453: Logic Colloquium. Symposium on Logic Held at Boston, 1972–73. Edited by R. Parikh. IV, 251 pages. 1975. DM 25,–

Vol. 454: J. Hirschfeld and W. H. Wheeler, Forcing, Arithmetic, Division Rings. VII, 266 pages. 1975. DM 25,–

Vol. 455: H. Kraft, Kommutative algebraische Gruppen und Ringe. III, 163 Seiten. 1975. DM 20,–

Vol. 456: R. M. Fossum, P. A. Griffith, and I. Reiten, Trivial Extensions of Abelian Categories. Homological Algebra of Trivial Extensions of Abelian Categories with Applications to Ring Theory. XI, 122 pages. 1975. DM 18,–

Vol. 457: Fractional Calculus and Its Applications. Proceedings of the International Conference Held at the University of New Haven, June 1974. Edited by B. Ross. VI, 381 pages. 1975. DM 32,–

Vol. 458: P. Walters, Ergodic Theory – Introductory Lectures. VI, 198 pages. 1975. DM 23,–

Vol. 459: Fourier Integral Operators and Partial Differential Equations. Proceedings 1974. Edited by J. Chazarain. VI, 372 pages. 1975. DM 32,–

Vol. 460: O. Loos, Jordan Pairs. XVI, 218 pages. 1975. DM 23,–

Vol. 461: Computational Mechanics. Proceedings 1974. Edited by J. T. Oden. VII, 328 pages. 1975. DM 30,–

Vol. 462: P. Gérardin, Construction de Séries Discrètes p-adiques. »Sur les séries discrètes non ramifiées des groupes réductifs déployés p-adiques«. III, 180 pages. 1975. DM 20,–

Vol. 463: H.-H. Kuo, Gaussian Measures in Banach Spaces. VI, 224 pages. 1975. DM 23,–

Vol. 464: C. Rockland, Hypoellipticity and Eigenvalue Asymptotics. III, 171 pages. 1975. DM 20,–

Vol. 465: Séminaire de Probabilités IX. Edited by P. A. Meyer. IV, 589 pages. 1975. DM 44,–

Vol. 466: Non-Commutative Harmonic Analysis. Proceedings 1974. Edited by J. Carmona, J. Dixmier and M. Vergne. VI, 231 pages. 1975. DM 23,–

Vol. 467: M. R. Essén, The Cos $\pi\lambda$ Theorem. VII, 112 pages. 1975. DM 18,–

Vol. 468: Dynamical Systems – Warwick 1974. Edited by A. K. Manning. X, 405 pages. 1975. DM 35,–

Vol. 469: E. Binz, Continuous Convergence on C(X). IX, 140 pages. 1975. DM 18,–

Vol. 470: R. Bowen, Equilibrium States and The Ergodic Theory of Anosov Diffeomorphisms. IV, 108 pages. 1975. DM 18,–

Vol. 471: R. S. Hamilton, Harmonic Maps of Manifolds with Boundary. IV, 168 pages. 1975. DM 20,–